ANTENNA DATA REFERENCE MANUAL
–including dimension tables

Other TAB books by the author:

No. 660 *FM Stereo-Quad Receiver Servicing Manual*
No. 787 *OP AMP Circuit Design & Applications*
No. 901 *CET License Handbook–2nd Edition*
No. 930 *Servicing Medical & Bioelectronic Equipment*
No. 1012 *How to Design and Build Electronic Instrumentation*
No. 1070 *Digital Interfacing with an Analog World*

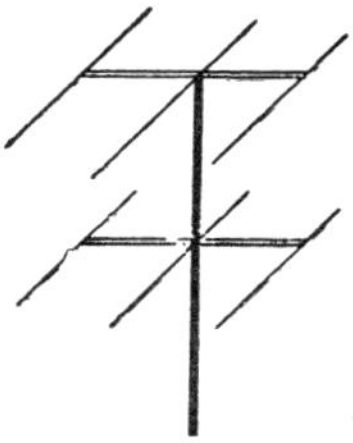

ANTENNA DATA REFERENCE MANUAL
—including dimension tables

by Joseph J. Carr, K4IPV

TAB BOOKS
BLUE RIDGE SUMMIT, PA. 17214

FIRST EDITION

FIRST PRINTING— OCTOBER 1979

Printed in the United States of America

Library of Congress Cataloging in Publication Data

Carr, Joseph J.
Antenna data reference manual, including dimension tables.

Includes index.
1. Antennas (Electronics)—Amateurs' manuals.
I. Title.
TK9956.C35 621.38'0283 79-16886
ISBN 0-8306-9738-1
ISBN 0-8306-1152-5 pbk.

Cover photo courtesy of National Cable Television Association.

Contents

Introduction

Here is a practical book, designed to be stuffed inside your antenna-building tool box, along with a heavy duty soldering iron, pliers, wire cutters, first aid supplies, an old liverwurst sandwich and an "807" or two. If you are too young to know what an "807" is, then let me tell you: it is a vacuum tube that resembled a beer bottle, so people call beer "807s." This book will not contain a lot of heavy theory. In fact, in many places we have taken the usual "amateur license" with theory, and made a few assumptions. This is especially true in the cases of the beam and quad antennas. In those cases, certain assumptions as to the "standard" antenna configuration were made . . . and the calculations reflect how I would have done it. If you do not like that way, then go look up a good text on beam antenna theory.

The purpose of this book is to let you get the antenna working with a minimum effort. The idea is to get on the air, not see how much neat "smoke" we know about antennas. Once on the air, then you can talk as much antenna theory as you like. But knowing a lot of smoke, and having no antenna, doesn't seem to practical.

Certain computer programs are given in this book. If you have a microcomputer that supports BASIC, fine. If not, don't worry about it. Except for the antenna bearings program of Chapter 14, most of the critical values for antenna lengths have been calculated for you.

The charts of this book suffer from the common calculator and computer problem of too many digits. We don't seriously expect you to cut an antenna to 63.23 feet . . . 62.2 feet is close enough.

We have concentrated mostly on amateur radio antennas, because hams are the ones who normally construct the largest number of homebrew antennas. But also covered, are citizens band antennas, shortwave listener antennas, and the antennas for the broadcast band, including transmitting antennas. More and more shortwave listeners have decided to build antennas customized for the various international broadcast bands, and some have even created rotating beam antennas. Even stationary beams are known, however, because of their ability to kill interference, such as Radio Moscow and others.

Joseph J. Carr, K4IPV

Chapter 1
Half-Wavelength Dipoles

The half-wavelength dipole is one of the most popular amateur radio antennas, especially among beginners. It is easy to build and easy to get working. Some of the more complex antennas, such as the beam antennas, will produce superior results, but are both electrically and mechanically more difficult to get working properly. Very few antennas will work so nicely with so little work as the ordinary half-wavelength dipole.

And work well it does! It is not impossible to work a lot of DX using little more than a dipole and a few dozen watts of rf power. In my own station, I have never been able to afford a power amplifier in the kilowatt class. Nor have I been able to afford a multielement beam antenna and all of the mechanical garbage that goes along with a beam. In my amateur career, my high-frequency operation has been with less than 200 watts and a series of half-wavelength dipoles.

The basic half-wavelength radiator (Fig. 1-1) is mounted horizontally to the earth, and is fed at the center. Each leg of the antenna is one-quarter wavelength. It is relatively important to cut these sections correctly. The idea is to make the length such that an electron fed at the center will travel out to the end, and back to the center in one-half of a cycle of the transmitter signal.

The graph of voltage and current as a function of length is shown in Fig. 1-1. Note that the voltage is maximum at the ends and minimum at the feedpoint. Similarly, the current at the feedpoint is maximum. This means that the antenna impedance is minimum at the feedpoint.

What is the impedance of a half wavelength dipole? Ask that question of 100 amateurs and at least 98 of them will tell you something between 70 and 75 ohms! But the truth is that the feedpoint impedance is 72 ohms, *only* in free space, i.e. at a location that is many wavelengths from any conducting object, including the earth's surface. The actual impedance of a real dipole, mounted close to the earth's surface, will vary from about 40 to 130 ohms. A

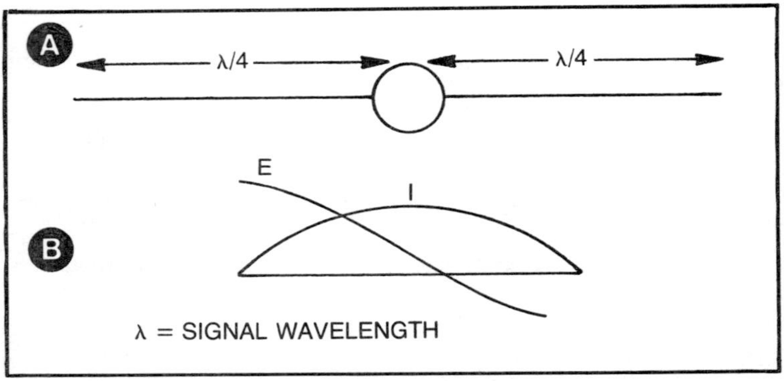

Fig. 1-1. Half-wavelength dipole at (A) and current (I)-voltage (E) distribution at (B).

discussion of this phenomenon is given by Rufus Turner on page 58 of his book, TAB book No. 829, *Impedance.*

The radiation pattern for a half-wavelength dipole is shown in Fig. 1-2. The ordinary dipole is bidirectional; there are two major radiation lobes at right angles to the wire radiator. The radiation off the ends of the dipole is near zero. Note, however, that real antennas, located in most neighborhoods, will not have such sharply defined lobes. This accounts for many amateurs having worked DX stations off the ends of their antennas.

Most amateurs, especially Novices, are familiar with the fact that a half-wavelength 40-meter dipole works nicely on the 15-meter band. But at 15 meters the antenna is a 3/2-wavelength radiator, so the radiation pattern is different from the half-wavelength case. The pattern for an ideal 3/2-wavelength radiator is shown in Fig. 1-3.

Figure 1-4 shows the construction for a half-wavelength dipole antenna. The center insulator may be a simple egg or glass insulator, or may be a special (and often costly) center insulator. Both HyGain

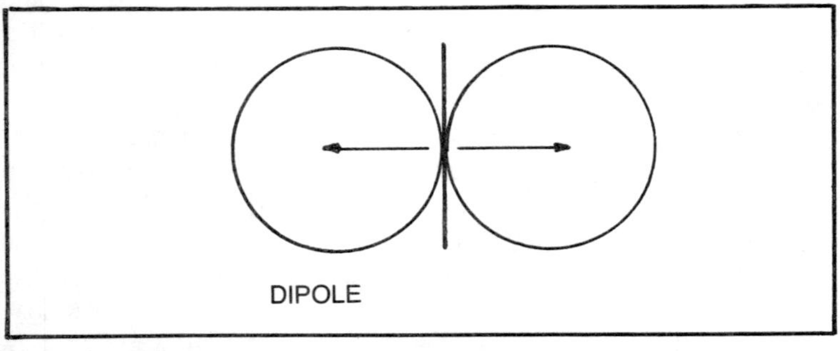

Fig. 1-2. Directional pattern (view from above) of a half-wave dipole.

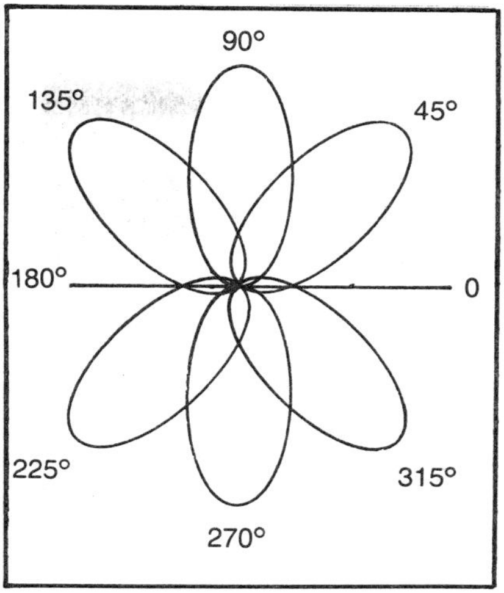

Fig. 1-3. Polar pattern of ¾-wavelength dipole.

and B&W have manufactured these devices. In my own case, I have used almost all of the commercial center insulators, as well as most forms of regular insulator. In fact, once I even used an old tooth brush as the center insulator!

Each half of the radiator is cut from a piece of wire. Various types of wire may be used for this. For strength, the best wire is copper-clad, steel wire, often referred to by its trade name, Copperweld. But copper-clad, steel wire is both expensive and scarce. Recently, I tried to buy such wire in the Washington, D.C. area and met with no luck at the amateur and commercial suppliers of electronic parts! All of them said they could order it if I wanted a sufficiently large quantity (enough to supply my local radio club for a year!).

The next best wire is ordinary stranded, copper, wire. In most cases, No. 14 wire is considered best. It is a good compromise between the large diameter needed for wide bandwidth, and the

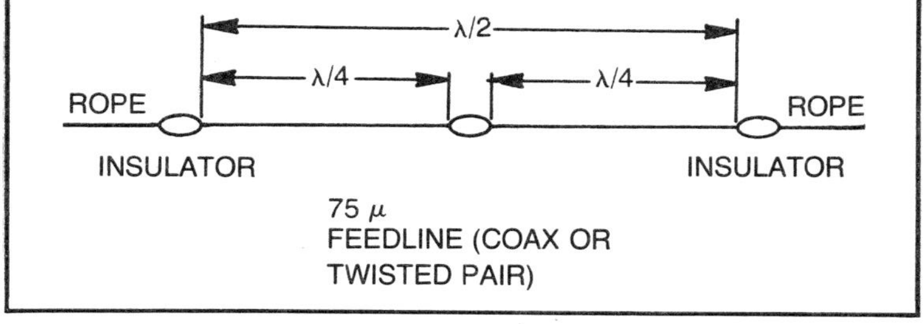

Fig. 1-4. Construction of a wire dipole.

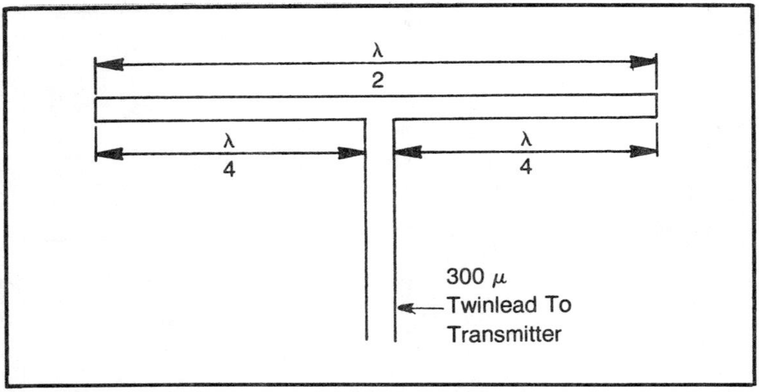

Fig. 1-5. Construction of a folded dipole.

small diameter needed for low weight. Remember, a 125-foot piece of wire (as used in 80-meter antennas) may weigh enough to break itself under its own weight. And when wind hits the structure, then goodbye antenna!

Solid copper wire usually fatigues, and breaks, in a short period of time. This is why stranded wire is preferred.

An example of the *folded dipole* antenna is shown in Fig. 1-5. This type of antenna uses two parallel radiators. One is continuous, while the other is broken at the center to accommodate the feedpoint. The nominal feedpoint impedance of this type of antenna is 300 ohms, so ordinary television antenna twin-lead transmission line can be used to feed the folded dipole. Note that 300-ohm twin-lead is also used for the two radiators.

DIPOLE LENGTHS

Both types of simple, half-wavelength, dipoles are the same physical length. The formula for a perfect, ideal, half-wavelength antenna is (in free space):

$$L = \frac{492}{F}$$

where L is the length of the radiator in *feet* and F is the frequency in megahertz.

☐ **Example**

Calculate the length in feet of a half-wavelength dipole cut for a resonant frequency of 6000 kHz. Assume the antenna is in "free space."

$$L = 492/F$$
$$L = 492/6$$
$$L = 82 \text{ feet}$$

Of course, the length of each section of the radiator will be exactly one-half of this length, or 41 feet.

For practical antennas—those close to the earth's surface—we must use a modified formula. This is due to phenomena like *end effect*. For all practical dipoles—close to the earth's surface or other conductive objects—the length is given by:

$$L = \frac{468}{F}$$

where L is the length in feet and F is the frequency in megahertz (MHz).

☐ **Example**

Calculate the length of the 6 MHz antenna in the previous example, if the antenna is to be installed less than "many wavelengths" from the earth's surface.

$$L = 468/F$$
$$L = 468/6$$
$$L = 78 \text{ feet}$$

(the length of the individual sections is one-half of this, or 39 feet)

TABLES

Tables giving the lengths for *practical* half-wavelength dipole antennas at popular frequencies are given in the section to follow.

TABLE 1-1
HALF WAVELENGTH DIPOLE
300 - 500 KILOHERTZ
USE: MARINE/RADIO NAVIGATION

Measurements, in feet, advance in 1 kHz increments from left to right across columns.

kHz	feet	feet	feet	feet	feet	feet	feet	feet	feet	feet
300	1560	1554.81	1549.66	1544.55	1539.47	1534.42	1529.41	1524.42	1519.48	1514.56
310	1509.67	1504.82	1500	1495.2	1490.44	1485.71	1481.01	1476.34	1471.69	1467.08
320	1462.5	1457.94	1453.41	1448.91	1444.44	1440	1435.58	1431.19	1426.82	1422.49
330	1418.18	1413.89	1409.63	1405.4	1401.19	1397.01	1392.85	1388.72	1384.61	1380.53
340	1376.47	1372.43	1368.42	1364.43	1360.46	1356.52	1352.6	1348.7	1344.82	1340.97
350	1337.14	1333.33	1329.54	1325.77	1322.03	1318.3	1314.6	1310.92	1307.26	1303.62
360	1300	1296.39	1292.81	1289.25	1285.71	1282.19	1278.68	1275.2	1271.73	1268.29
370	1264.86	1261.45	1258.06	1254.69	1251.33	1248	1244.68	1241.37	1238.09	1234.82
380	1231.57	1228.34	1225.13	1221.93	1218.75	1215.58	1212.43	1209.3	1206.18	1203.08
390	1200	1196.93	1193.87	1190.83	1187.81	1184.81	1181.81	1178.84	1175.87	1172.93
400	1170	1167.08	1164.17	1161.29	1158.41	1155.55	1152.7	1149.87	1147.05	1144.25
410	1141.46	1138.68	1135.92	1133.17	1130.43	1127.71	1125	1122.3	1119.61	1116.94
420	1114.28	1111.63	1109	1106.38	1103.77	1101.17	1098.59	1096.01	1093.45	1090.9
430	1088.37	1085.84	1083.33	1080.83	1078.34	1075.86	1073.39	1070.93	1068.49	1066.05
440	1063.63	1061.22	1058.82	1056.43	1054.05	1051.68	1049.32	1046.97	1044.64	1042.31
450	1040	1037.69	1035.39	1033.11	1030.83	1028.57	1026.31	1024.07	1021.83	1019.6
460	1017.39	1015.18	1012.98	1010.79	1008.62	1006.45	1004.29	1002.14		997.86
470	995.74	993.63	991.52	989.42	987.34	985.26	983.19	981.13	979.07	977.03
480	975	972.97	970.95	968.94	966.94	964.94	962.96	960.98	959.01	957.05
490	955.1	953.15	951.21	949.29	947.36	945.45	943.54	941.64	939.75	937.87
500	936	934.13	932.27	930.41	928.57	926.73	924.9	923.07	921.25	919.44

TABLE 1-2
HALF WAVELENGTH DIPOLE
540 - 1605 KILOHERTZ
USE: AM BROADCASTING

Measurements, in feet, advance in 1 kHz increments from left to right across columns.

kHz	feet	feet	feet	feet	feet	feet	feet	feet	feet	feet
540	866.66	858.71	850.9	843.24	835.71	828.31	821.05	813.91	806.89	800
590	793.22	786.55	780	773.55	767.21	760.97	754.83	748.8	742.85	737
640	731.25	725.58	720	714.5	709.09	703.75	698.5	693.33	688.23	683.21
690	678.26	673.38	668.57	663.82	659.15	654.54	650	645.51	641.09	636.73
740	632.43	628.18	624	619.86	615.78	611.76	607.79	603.87	600	596.17
790	592.4	588.67	585	581.36	577.77	574.23	570.73	567.27	563.85	560.47
840	557.14	553.84	550.58	547.36	544.18	541.04	537.93	534.85	531.81	528.81
890	525.84	522.9	520	517.12	514.28	511.47	508.69	505.94	503.22	500.53
940	497.87	495.23	492.63	490.05	487.5	484.97	482.47	480	477.55	475.12
990	472.72	470.35	468	465.67	463.36	461.08	458.82	456.58	454.36	452.17
1040	450	447.84	445.71	443.6	441.5	439.43	437.38	435.34	433.33	431.33
1090	429.35	427.39	425.45	423.52	421.62	419.73	417.85	416	414.15	412.33
1140	410.52	408.73	406.95	405.19	403.44	401.71	400	398.29	396.61	394.93
1190	393.27	391.63	390	388.38	386.77	385.18	383.6	382.04	380.48	378.94
1240	377.41	375.9	374.4	372.9	371.42	369.96	368.5	367.05	365.62	364.2
1290	362.79	361.38	360	358.62	357.25	355.89	354.54	353.2	351.87	350.56
1340	349.25	347.95	346.66	345.38	344.11	342.85	341.6	340.36	339.13	337.9
1390	336.69	335.48	334.28	333.09	331.91	330.74	329.57	328.42	327.27	326.13
1440	325	323.87	322.75	321.64	320.54	319.45	318.36	317.28	316.21	315.15
1490	314.09	313.04	312	310.96	309.93	308.91	307.89	306.88	305.88	304.88
1540	303.89	302.91	301.93	300.96	300	299.04	298.08	297.14	296.2	295.26
1590	294.33	293.41	292.5	291.58	290.68	289.78	288.88	288	287.11	286.23

TABLE 1-3
HALF WAVELENGTH DIPOLE
1800 - 2000 KILOHERTZ
USE: AMATEUR

Measurements, in feet, advance in 1 kHz increments from left to right across columns.

kHz	feet	feet	feet	feet	feet	feet	feet	feet	feet	feet
1800	260	259.85	259.71	259.56	259.42	259.27	259.13	258.99	258.84	258.7
1810	258.56	258.42	258.27	258.13	257.99	257.85	257.7	257.56	257.42	257.28
1820	257.14	257	256.86	256.71	256.57	256.43	256.29	256.15	256.01	255.87
1830	255.73	255.59	255.45	255.31	255.17	255.04	254.9	254.76	254.62	254.48
1840	254.34	254.2	254.07	253.93	253.79	253.65	253.52	253.38	253.24	253.1
1850	252.97	252.83	252.69	252.56	252.42	252.29	252.15	252.01	251.88	251.74
1860	251.61	251.47	251.34	251.2	251.07	250.93	250.8	250.66	250.53	250.4
1870	250.26	250.13	250	249.86	249.73	249.6	249.46	249.33	249.2	249.06
1880	248.93	248.8	248.67	248.53	248.4	248.27	248.14	248.01	247.88	247.75
1890	247.61	247.48	247.35	247.22	247.09	246.96	246.83	246.7	246.57	246.44
1900	246.31	246.18	246.05	245.92	245.79	245.66	245.54	245.41	245.28	245.15
1910	245.02	244.89	244.76	244.64	244.51	244.38	244.25	244.13	244	243.87
1920	243.75	243.62	243.49	243.36	243.24	243.11	242.99	242.86	242.73	242.61
1930	242.48	242.36	242.23	242.11	241.98	241.86	241.73	241.61	241.48	241.36
1940	241.23	241.11	240.98	240.86	240.74	240.61	240.49	240.36	240.24	240.12
1950	240	239.87	239.75	239.63	239.5	239.38	239.26	239.14	239.01	238.89
1960	238.77	238.65	238.53	238.41	238.28	238.16	238.04	237.92	237.8	237.68
1970	237.56	237.44	237.32	237.2	237.08	236.96	236.84	236.72	236.6	236.48
1980	236.36	236.24	236.12	236	235.88	235.76	235.64	235.53	235.41	235.29
1990	235.17	235.05	234.93	234.82	234.7	234.58	234.46	234.35	234.23	234.11
2000	234	233.88	233.76	233.64	233.53	233.41	233.3	233.18	233.06	232.95

TABLE 1-4
HALF-WAVELENGTH DIPOLE
3500 - 4000 KILOHERTZ
USE: AMATEUR

Measurements, in feet, advance in 1 kHz increments from left to right across columns.

kHz	feet	feet	feet	feet	feet	feet	feet	feet	feet	feet
3500	133.71	133.63	133.56	133.48	133.4	133.33	133.25	133.18	133.1	133.03
3520	132.95	132.87	132.8	132.72	132.65	132.57	132.5	132.42	132.35	132.27
3540	132.2	132.12	132.05	131.97	131.9	131.83	131.75	131.68	131.6	131.53
3560	131.46	131.38	131.31	131.23	131.16	131.09	131.01	130.94	130.87	130.79
3580	130.72	130.65	130.58	130.5	130.43	130.36	130.28	130.21	130.14	130.07
3600	130	129.92	129.85	129.78	129.71	129.63	129.56	129.49	129.42	129.35
3620	129.28	129.21	129.13	129.06	128.99	128.92	128.85	128.78	128.71	128.64
3640	128.57	128.5	128.43	128.35	128.28	128.21	128.14	128.07	128	127.93
3660	127.86	127.79	127.72	127.65	127.58	127.52	127.45	127.38	127.31	127.24
3680	127.17	127.1	127.03	126.96	126.89	126.82	126.76	126.69	126.62	126.55
3700	126.48	126.41	126.34	126.28	126.21	126.14	126.07	126	125.94	125.87
3720	125.8	125.73	125.67	125.6	125.53	125.46	125.4	125.33	125.26	125.2
3740	125.13	125.06	125	124.93	124.86	124.8	124.73	124.66	124.6	124.53
3760	124.46	124.4	124.33	124.26	124.2	124.13	124.07	124	123.94	123.87
3780	123.8	123.74	123.67	123.61	123.54	123.48	123.41	123.35	123.28	123.22
3800	123.15	123.09	123.02	122.96	122.89	122.83	122.77	122.7	122.64	122.57
3820	122.51	122.44	122.38	122.32	122.25	122.19	122.12	122.06	122	121.93
3840	121.87	121.81	121.74	121.68	121.62	121.55	121.49	121.43	121.36	121.3
3860	121.24	121.18	121.11	121.05	120.99	120.93	120.86	120.8	120.74	120.68
3880	120.61	120.55	120.49	120.43	120.37	120.3	120.24	120.18	120.12	120.06
3900	120	119.93	119.87	119.81	119.75	119.69	119.63	119.57	119.5	119.44
3920	119.38	119.32	119.26	119.2	119.14	119.08	119.02	118.96	118.9	118.84
3940	118.78	118.72	118.66	118.6	118.54	118.48	118.42	118.36	118.3	118.24
3960	118.18	118.12	118.06	118	117.94	117.88	117.82	117.76	117.7	117.64
3980	117.58	117.52	117.46	117.41	117.35	117.29	117.23	117.17	117.11	117.05
4000	117	116.94	116.88	116.82	116.76	116.7	116.65	116.59	116.53	116.47

TABLE 1-5
HALF-WAVELENGTH DIPOLE
5500 - 6000 KILOHERTZ
USE: INT'L. BROADCAST

Measurements, in feet, advance in 1 kHz increments from left to right across columns.

kHz	feet	reet	feet	feet	feet	feet	feet	feet	feet	feet
5500	85.09	85.05	85.02	84.99	84.96	84.93	84.9	84.87	84.84	84.81
5520	84.78	84.75	84.72	84.69	84.65	84.62	84.59	84.56	84.53	84.5
5540	84.47	84.44	84.41	84.38	84.35	84.32	84.29	84.26	84.23	84.2
5560	84.17	84.14	84.11	84.08	84.05	84.02	83.99	83.96	83.93	83.9
5580	83.87	83.84	83.81	83.78	83.75	83.72	83.69	83.66	83.63	83.6
5600	83.57	83.54	83.51	83.48	83.45	83.42	83.39	83.36	83.33	83.3
5620	83.27	83.24	83.21	83.18	83.15	83.12	83.09	83.06	83.03	83
5640	82.97	82.94	82.91	82.89	82.86	82.83	82.8	82.77	82.74	82.71
5660	82.68	82.65	82.62	82.59	82.56	82.53	82.51	82.48	82.45	82.42
5680	82.39	82.36	82.33	82.3	82.27	82.24	82.22	82.19	82.16	82.13
5700	82.1	82.07	82.04	82.01	81.99	81.96	81.93	81.9	81.87	81.84
5720	81.81	81.78	81.76	81.73	81.7	81.67	81.64	81.61	81.58	81.56
5740	81.53	81.5	81.47	81.44	81.41	81.39	81.36	81.33	81.3	81.27
5760	81.25	81.22	81.19	81.16	81.13	81.1	81.08	81.05	81.02	80.99
5780	80.96	80.94	80.91	80.88	80.85	80.82	80.8	80.77	80.74	80.71
5800	80.68	80.66	80.63	80.6	80.57	80.55	80.52	80.49	80.46	80.44
5820	80.41	80.38	80.35	80.32	80.3	80.27	80.24	80.21	80.19	80.16
5840	80.13	80.1	80.08	80.05	80.02	80	79.97	79.94	79.91	79.89
5860	79.86	79.83	79.8	79.78	79.75	79.72	79.7	79.67	79.64	79.61
5880	79.59	79.56	79.53	79.51	79.48	79.45	79.42	79.4	79.37	79.34
5900	79.32	79.29	79.26	79.24	79.21	79.18	79.16	79.13	79.1	79.08
5920	79.05	79.02	79	78.97	78.94	78.92	78.89	78.86	78.84	78.81
5940	78.78	78.76	78.73	78.7	78.68	78.65	78.62	78.6	78.57	78.54
5960	78.52	78.49	78.47	78.44	78.41	78.39	78.36	78.33	78.31	78.28
5980	78.26	78.23	78.2	78.18	78.15	78.13	78.1	78.07	78.05	78.02
6000	78	77.97	77.94	77.92	77.89	77.87	77.84	77.81	77.79	77.76

TABLE 1-6
HALF-WAVELENGTH DIPOLE
6000 - 6500 KILOHERTZ
USE: INT'L. BROADCAST

Measurements, in feet, advance in 1 kHz increments from left to right across columns.

kHz	feet	feet	feet	feet	feet	feet	feet	feet	feet	feet
6000	78	77.97	77.94	77.92	77.89	77.87	77.84	77.81	77.79	77.76
6020	77.74	77.71	77.68	77.66	77.63	77.61	77.58	77.56	77.53	77.5
6040	77.48	77.45	77.43	77.4	77.38	77.35	77.32	77.3	77.27	77.25
6060	77.22	77.2	77.17	77.15	77.12	77.1	77.07	77.04	77.02	76.99
6080	76.97	76.94	76.92	76.89	76.87	76.84	76.82	76.79	76.77	76.74
6100	76.72	76.69	76.67	76.64	76.62	76.59	76.57	76.54	76.52	76.49
6120	76.47	76.44	76.42	76.39	76.37	76.34	76.32	76.29	76.27	76.24
6140	76.22	76.19	76.17	76.14	76.12	76.09	76.07	76.04	76.02	75.99
6160	75.97	75.94	75.92	75.9	75.87	75.85	75.82	75.8	75.77	75.75
6180	75.72	75.7	75.67	75.65	75.63	75.6	75.58	75.55	75.53	75.5
6200	75.48	75.45	75.43	75.41	75.38	75.36	75.33	75.31	75.28	75.26
6220	75.24	75.21	75.19	75.16	75.14	75.12	75.09	75.07	75.04	75.02
6240	75	74.97	74.95	74.92	74.9	74.88	74.85	74.83	74.8	74.78
6260	74.76	74.73	74.71	74.68	74.66	74.64	74.61	74.59	74.56	74.54
6280	74.52	74.49	74.47	74.45	74.42	74.4	74.38	74.35	74.33	74.3
6300	74.28	74.26	74.23	74.21	74.19	74.16	74.14	74.12	74.09	74.07
6320	74.05	74.02	74	73.98	73.95	73.93	73.91	73.88	73.86	73.84
6340	73.81	73.79	73.77	73.74	73.72	73.7	73.67	73.65	73.63	73.6
6360	73.58	73.56	73.53	73.51	73.49	73.46	73.44	73.42	73.4	73.37
6380	73.35	73.33	73.3	73.28	73.26	73.23	73.21	73.19	73.17	73.14
6400	73.12	73.1	73.07	73.05	73.03	73.01	72.98	72.96	72.94	72.91
6420	72.89	72.87	72.85	72.82	72.8	72.78	72.76	72.73	72.71	72.69
6440	72.67	72.64	72.62	72.6	72.58	72.55	72.53	72.51	72.49	72.46
6460	72.44	72.42	72.4	72.37	72.35	72.33	72.31	72.28	72.26	72.24
6480	72.22	72.19	72.17	72.15	72.13	72.11	72.08	72.06	72.04	72.02
6500	72	71.97	71.95	71.93	71.91	71.88	71.86	71.84	71.82	71.8

TABLE 1-7
HALF-WAVELENGTH DIPOLE
7000 - 7300 KILOHERTZ
USE: AMATEUR

Measurements, in feet, advance in 1 kHz increments from left to right across columns.

kHz	feet	feet	feet	feet	feet	feet	feet	feet	feet	feet
7000	66.85	66.83	66.81	66.79	66.78	66.76	66.74	66.72	66.7	66.68
7020	66.66	66.64	66.62	66.6	66.59	66.57	66.55	66.53	66.51	66.49
7040	66.47	66.45	66.43	66.42	66.4	66.38	66.36	66.34	66.32	66.3
7060	66.28	66.27	66.25	66.23	66.21	66.19	66.17	66.15	66.13	66.12
7080	66.1	66.08	66.06	66.04	66.02	66	65.98	65.97	65.95	65.93
7100	65.91	65.89	65.87	65.85	65.84	65.82	65.8	65.78	65.76	65.74
7120	65.73	65.71	65.69	65.67	65.65	65.63	65.61	65.6	65.58	65.56
7140	65.54	65.52	65.5	65.49	65.47	65.45	65.43	65.41	65.39	65.38
7160	65.36	65.34	65.32	65.3	65.29	65.27	65.25	65.23	65.21	65.19
7180	65.18	65.16	65.14	65.12	65.1	65.09	65.07	65.05	65.03	65.01
7200	65	64.98	64.96	64.94	64.92	64.9	64.89	64.87	64.85	64.83
7220	64.81	64.8	64.78	64.76	64.74	64.73	64.71	64.69	64.67	64.65
7240	64.64	64.62	64.6	64.58	64.56	64.55	64.53	64.51	64.49	64.48
7260	64.46	64.44	64.42	64.4	64.39	64.37	64.35	64.33	64.32	64.3
7280	64.28	64.26	64.25	64.23	64.21	64.19	64.17	64.16	64.14	64.12
7300	64.1	64.09	64.07	64.05	64.03	64.02	64	63.98	63.96	63.95

TABLE 1-8
HALF-WAVELENGTH DIPOLE
9500 - 10000 KILOHERTZ
USE: INT'L. BROADCAST

Measurements, in feet, advance in 1 kHz increments from left to right across columns.

kHz	feet	feet	feet	feet	feet	feet	feet	feet	feet	feet
9500	49.26	49.25	49.24	49.23	49.22	49.21	49.2	49.19	49.18	49.16
9520	49.15	49.14	49.13	49.12	49.11	49.1	49.09	49.08	49.07	49.06
9540	49.05	49.04	49.03	49.02	49.01	49	48.99	48.98	48.97	48.96
9560	48.95	48.94	48.93	48.92	48.91	48.9	48.89	48.88	48.87	48.86
9580	48.85	48.84	48.83	48.82	48.81	48.8	48.79	48.78	48.77	48.76
9600	48.75	48.73	48.72	48.71	48.7	48.69	48.68	48.67	48.66	48.65
9620	48.64	48.63	48.62	48.61	48.6	48.59	48.58	48.57	48.56	48.55
9640	48.54	48.53	48.52	48.51	48.5	48.49	48.48	48.47	48.46	48.45
9660	48.44	48.43	48.42	48.41	48.4	48.39	48.38	48.37	48.36	48.35
9680	48.34	48.33	48.32	48.31	48.3	48.29	48.28	48.27	48.26	48.25
9700	48.24	48.23	48.22	48.21	48.2	48.19	48.18	48.17	48.16	48.15
9720	48.14	48.13	48.12	48.11	48.1	48.09	48.08	48.07	48.06	48.05
9740	48.04	48.03	48.02	48.01	48	48	47.99	47.98	47.97	47.96
9760	47.95	47.94	47.93	47.92	47.91	47.9	47.89	47.88	47.87	47.86
9780	47.85	47.84	47.83	47.82	47.81	47.8	47.79	47.78	47.77	47.76
9800	47.75	47.74	47.73	47.72	47.71	47.7	47.69	47.68	47.67	47.66
9820	47.65	47.64	47.63	47.62	47.61	47.6	47.59	47.58	47.58	47.57
9840	47.56	47.55	47.54	47.53	47.52	47.51	47.5	47.49	47.48	47.47
9860	47.46	47.45	47.44	47.43	47.42	47.41	47.4	47.39	47.38	47.37
9880	47.36	47.35	47.34	47.33	47.33	47.32	47.31	47.3	47.29	47.28
9900	47.27	47.26	47.25	47.24	47.23	47.22	47.21	47.2	47.19	47.18
9920	47.17	47.16	47.15	47.14	47.13	47.12	47.12	47.11	47.1	47.09
9940	47.08	47.07	47.06	47.05	47.04	47.03	47.02	47.01	47	46.99
9960	46.98	46.97	46.96	46.95	46.95	46.94	46.93	46.92	46.91	46.9
9980	46.89	46.88	46.87	46.86	46.85	46.84	46.83	46.82	46.81	46.8
10000	46.8	46.79	46.78	46.77	46.76	46.75	46.74	46.73	46.72	46.71

TABLE 1-9
HALF-WAVELENGTH DIPOLE
14000 - 14450 KILOHERTZ
USE: AMATEUR

Measurements, in feet, advance in 1 kHz increments from left to right across columns.

kHz	feet	feet	feet	feet	feet	feet	feet	feet	feet	feet
14000	33.42	33.42	33.41	33.41	33.4	33.4	33.39	33.39	33.39	33.38
14020	33.38	33.37	33.37	33.36	33.36	33.35	33.35	33.34	33.34	33.33
14040	33.33	33.32	33.32	33.31	33.31	33.3	33.3	33.3	33.29	33.29
14060	33.28	33.28	33.27	33.27	33.26	33.26	33.25	33.25	33.24	33.24
14080	33.23	33.23	33.22	33.22	33.21	33.21	33.21	33.2	33.2	33.19
14100	33.19	33.18	33.18	33.17	33.17	33.16	33.16	33.15	33.15	33.14
14120	33.14	33.13	33.13	33.13	33.12	33.12	33.11	33.11	33.1	33.1
14140	33.09	33.09	33.08	33.08	33.07	33.07	33.06	33.06	33.06	33.05
14160	33.05	33.04	33.04	33.03	33.03	33.02	33.02	33.01	33.01	33
14180	33	32.99	32.99	32.99	32.98	32.98	32.97	32.97	32.96	32.96
14200	32.95	32.95	32.94	32.94	32.93	32.93	32.92	32.92	32.92	32.91
14220	32.91	32.9	32.9	32.89	32.89	32.88	32.88	32.87	32.87	32.86
14240	32.86	32.86	32.85	32.85	32.84	32.84	32.83	32.83	32.82	32.82
14260	32.81	32.81	32.8	32.8	32.8	32.79	32.79	32.78	32.78	32.77
14280	32.77	32.76	32.76	32.75	32.75	32.75	32.74	32.74	32.73	32.73
14300	32.72	32.72	32.71	32.71	32.7	32.7	32.69	32.69	32.69	32.68
14320	32.68	32.67	32.67	32.66	32.66	32.65	32.65	32.64	32.64	32.64
14340	32.63	32.63	32.62	32.62	32.61	32.61	32.6	32.6	32.59	32.59
14360	32.59	32.58	32.58	32.57	32.57	32.56	32.56	32.55	32.55	32.54
14380	32.54	32.54	32.53	32.53	32.52	32.52	32.51	32.51	32.5	32.5
14400	32.5	32.49	32.49	32.48	32.48	32.47	32.47	32.46	32.46	32.45
14420	32.45	32.45	32.44	32.44	32.43	32.43	32.42	32.42	32.41	32.41
14440	32.4	32.4	32.4	32.39	32.39	32.38	32.38	32.37	32.37	32.36

TABLE 1-11
HALF-WAVELENGTH DIPOLE
15500 - 16000 KILOHERTZ
USE: INT'L. BROADCAST

Measurements, in feet, advance in 1 kHz increments from left to right across columns.

kHz	feet	feet	feet	feet	feet	feet	feet	feet	feet	feet
15500	30.19	30.18	30.18	30.18	30.17	30.17	30.17	30.16	30.16	30.15
15520	30.15	30.15	30.14	30.14	30.13	30.13	30.13	30.12	30.12	30.11
15540	30.11	30.11	30.1	30.1	30.1	30.09	30.09	30.08	30.08	30.08
15560	30.07	30.07	30.06	30.06	30.06	30.05	30.05	30.05	30.04	30.04
15580	30.03	30.03	30.03	30.02	30.02	30.01	30.01	30.01	30	30
15600	30	29.99	29.99	29.98	29.98	29.98	29.97	29.97	29.96	29.96
15620	29.96	29.95	29.95	29.95	29.94	29.94	29.93	29.93	29.93	29.92
15640	29.92	29.91	29.91	29.91	29.9	29.9	29.9	29.89	29.89	29.88
15660	29.88	29.88	29.87	29.87	29.86	29.86	29.86	29.85	29.85	29.85
15680	29.84	29.84	29.83	29.83	29.83	29.82	29.82	29.82	29.81	29.81
15700	29.8	29.8	29.8	29.79	29.79	29.78	29.78	29.78	29.77	29.77
15720	29.77	29.76	29.76	29.75	29.75	29.75	29.74	29.74	29.74	29.73
15740	29.73	29.72	29.72	29.72	29.71	29.71	29.71	29.7	29.7	29.69
15760	29.69	29.69	29.68	29.68	29.68	29.67	29.67	29.66	29.66	29.66
15780	29.65	29.65	29.65	29.64	29.64	29.63	29.63	29.63	29.62	29.62
15800	29.62	29.61	29.61	29.6	29.6	29.6	29.59	29.59	29.59	29.58
15820	29.58	29.57	29.57	29.57	29.56	29.56	29.56	29.55	29.55	29.54
15840	29.54	29.54	29.53	29.53	29.53	29.52	29.52	29.51	29.51	29.51
15860	29.5	29.5	29.5	29.49	29.49	29.48	29.48	29.48	29.47	29.47
15880	29.47	29.46	29.46	29.45	29.45	29.45	29.44	29.44	29.44	29.43
15900	29.43	29.43	29.42	29.42	29.41	29.41	29.41	29.4	29.4	29.4
15920	29.39	29.39	29.38	29.38	29.38	29.37	29.37	29.37	29.36	29.36
15940	29.36	29.35	29.35	29.34	29.34	29.34	29.33	29.33	29.33	29.32
15960	29.32	29.31	29.31	29.31	29.3	29.3	29.3	29.29	29.29	29.29
15980	29.28	29.28	29.27	29.27	29.27	29.26	29.26	29.26	29.25	29.25
16000	29.25	29.24	29.24	29.23	29.23	29.23	29.22	29.22	29.22	29.21

TABLE 1-12
HALF-WAVELENGTH DIPOLE
16500 - 17000 KILOHERTZ
USE: INT'L. BROADCAST

Measurements, in feet, advance in 1 kHz increments from left to right across columns.

kHz	feet	feet	feet	feet	feet	feet	feet	feet	feet	feet
16500	28.36	28.36	28.35	28.35	28.34	28.34	28.34	28.33	28.33	28.33
16520	28.32	28.32	28.32	28.31	28.31	28.31	28.3	28.3	28.3	28.29
16540	28.29	28.29	28.28	28.28	28.28	28.27	28.27	28.27	28.26	28.26
16560	28.26	28.25	28.25	28.25	28.24	28.24	28.24	28.23	28.23	28.23
16580	28.22	28.22	28.21	28.21	28.21	28.2	28.2	28.2	28.19	28.19
16600	28.19	28.18	28.18	28.18	28.17	28.17	28.17	28.16	28.16	28.16
16620	28.15	28.15	28.15	28.14	28.14	28.14	28.13	28.13	28.13	28.12
16640	28.12	28.12	28.11	28.11	28.11	28.1	28.1	28.1	28.09	28.09
16660	28.09	28.08	28.08	28.08	28.07	28.07	28.07	28.06	28.06	28.06
16680	28.05	28.05	28.05	28.04	28.04	28.04	28.03	28.03	28.03	28.02
16700	28.02	28.02	28.01	28.01	28.01	28	28	28	27.99	27.99
16720	27.99	27.98	27.98	27.98	27.97	27.97	27.97	27.96	27.96	27.96
16740	27.95	27.95	27.95	27.94	27.94	27.94	27.93	27.93	27.93	27.92
16760	27.92	27.92	27.91	27.91	27.91	27.9	27.9	27.9	27.89	27.89
16780	27.89	27.88	27.88	27.88	27.87	27.87	27.87	27.86	27.86	27.86
16800	27.85	27.85	27.85	27.84	27.84	27.84	27.83	27.83	27.83	27.82
16820	27.82	27.82	27.81	27.81	27.81	27.8	27.8	27.8	27.79	27.79
16840	27.79	27.78	27.78	27.78	27.77	27.77	27.77	27.76	27.76	27.76
16860	27.75	27.75	27.75	27.74	27.74	27.74	27.73	27.73	27.73	27.72
16880	27.72	27.72	27.71	27.71	27.71	27.7	27.7	27.7	27.69	27.69
16900	27.69	27.68	27.68	27.68	27.67	27.67	27.67	27.66	27.66	27.66
16920	27.65	27.65	27.65	27.64	27.64	27.64	27.63	27.63	27.63	27.63
16940	27.62	27.62	27.62	27.61	27.61	27.61	27.6	27.6	27.6	27.59
16960	27.59	27.59	27.58	27.58	27.58	27.57	27.57	27.57	27.56	27.56
16980	27.56	27.55	27.55	27.55	27.54	27.54	27.54	27.53	27.53	27.53
17000	27.52	27.52	27.52	27.51	27.51	27.51	27.5	27.5	27.5	27.5

TABLE 1-13
HALF-WAVELENGTH DIPOLE
18500 - 19000 KILOHERTZ
USE: INT'L. BROADCAST

Measurements, in feet, advance in 1 kHz increments from left to right across columns.

kHz	feet	feet	feet	feet	feet	feet	feet	feet	feet	feet
18500	25.29	25.29	25.29	25.28	25.28	25.28	25.28	25.27	25.27	25.27
18520	25.26	25.26	25.26	25.26	25.25	25.25	25.25	25.25	25.24	25.24
18540	25.24	25.23	25.23	25.23	25.23	25.22	25.22	25.22	25.22	25.21
18560	25.21	25.21	25.21	25.2	25.2	25.2	25.19	25.19	25.19	25.19
18580	25.18	25.18	25.18	25.18	25.17	25.17	25.17	25.16	25.16	25.16
18600	25.16	25.15	25.15	25.15	25.15	25.14	25.14	25.14	25.13	25.13
18620	25.13	25.13	25.12	25.12	25.12	25.12	25.11	25.11	25.11	25.1
18640	25.1	25.1	25.1	25.09	25.09	25.09	25.09	25.08	25.08	25.08
18660	25.08	25.07	25.07	25.07	25.06	25.06	25.06	25.06	25.05	25.05
18680	25.05	25.05	25.04	25.04	25.04	25.04	25.03	25.03	25.03	25.02
18700	25.02	25.02	25.02	25.01	25.01	25.01	25.01	25	25	25
18720	25	24.99	24.99	24.99	24.98	24.98	24.98	24.98	24.97	24.97
18740	24.97	24.97	24.96	24.96	24.96	24.96	24.95	24.95	24.95	24.94
18760	24.94	24.94	24.94	24.93	24.93	24.93	24.93	24.92	24.92	24.92
18780	24.92	24.91	24.91	24.91	24.9	24.9	24.9	24.9	24.89	24.89
18800	24.89	24.89	24.88	24.88	24.88	24.88	24.87	24.87	24.87	24.86
18820	24.86	24.86	24.86	24.85	24.85	24.85	24.85	24.84	24.84	24.84
18840	24.84	24.83	24.83	24.83	24.83	24.82	24.82	24.82	24.81	24.81
18860	24.81	24.81	24.8	24.8	24.8	24.8	24.79	24.79	24.79	24.79
18880	24.78	24.78	24.78	24.78	24.77	24.77	24.77	24.76	24.76	24.76
18900	24.76	24.75	24.75	24.75	24.75	24.74	24.74	24.74	24.74	24.73
18920	24.73	24.73	24.73	24.72	24.72	24.72	24.72	24.71	24.71	24.71
18940	24.7	24.7	24.7	24.7	24.69	24.69	24.69	24.69	24.68	24.68
18960	24.68	24.68	24.67	24.67	24.67	24.67	24.66	24.66	24.66	24.66
18980	24.65	24.65	24.65	24.64	24.64	24.64	24.64	24.63	24.63	24.63
19000	24.63	24.62	24.62	24.62	24.62	24.61	24.61	24.61	24.61	24.6

TABLE 1-14
HALF-WAVELENGTH DIPOLE
21000 - 21450 KILOHERTZ
USE: AMATEUR

Measurements, in feet, advance in 1 kHz increments from left to right across columns.

kHz	feet	feet	feet	feet	feet	feet	feet	feet	feet	feet
21000	22.28	22.28	22.28	22.27	22.27	22.27	22.27	22.27	22.26	22.26
21020	22.26	22.26	22.26	22.25	22.25	22.25	22.25	22.24	22.24	22.24
21040	22.24	22.24	22.23	22.23	22.23	22.23	22.23	22.22	22.22	22.22
21060	22.22	22.22	22.21	22.21	22.21	22.21	22.2	22.2	22.2	22.2
21080	22.2	22.19	22.19	22.19	22.19	22.19	22.18	22.18	22.18	22.18
21100	22.18	22.17	22.17	22.17	22.17	22.16	22.16	22.16	22.16	22.16
21120	22.15	22.15	22.15	22.15	22.15	22.14	22.14	22.14	22.14	22.14
21140	22.13	22.13	22.13	22.13	22.12	22.12	22.12	22.12	22.12	22.11
21160	22.11	22.11	22.11	22.11	22.1	22.1	22.1	22.1	22.1	22.09
21180	22.09	22.09	22.09	22.09	22.08	22.08	22.08	22.08	22.07	22.07
21200	22.07	22.07	22.07	22.06	22.06	22.06	22.06	22.06	22.05	22.05
21220	22.05	22.05	22.05	22.04	22.04	22.04	22.04	22.04	22.03	22.03
21240	22.03	22.03	22.02	22.02	22.02	22.02	22.02	22.01	22.01	22.01
21260	22.01	22.01	22	22	22	22	22	21.99	21.99	21.99
21280	21.99	21.99	21.98	21.98	21.98	21.98	21.98	21.97	21.97	21.97
21300	21.97	21.96	21.96	21.96	21.96	21.96	21.95	21.95	21.95	21.95
21320	21.95	21.94	21.94	21.94	21.94	21.94	21.93	21.93	21.93	21.93
21340	21.93	21.92	21.92	21.92	21.92	21.92	21.91	21.91	21.91	21.91
21360	21.91	21.9	21.9	21.9	21.9	21.89	21.89	21.89	21.89	21.89
21380	21.88	21.88	21.88	21.88	21.88	21.87	21.87	21.87	21.87	21.87
21400	21.86	21.86	21.86	21.86	21.86	21.85	21.85	21.85	21.85	21.85
21420	21.84	21.84	21.84	21.84	21.84	21.83	21.83	21.83	21.83	21.83
21440	21.82	21.82	21.82	21.82	21.82	21.81	21.81	21.81	21.81	21.81

TABLE 1-15
HALF-WAVELENGTH DIPOLE
25500 - 26000 KILOHERTZ
USE: INT'L. BROADCAST

Measurements, in feet, advance in 1 kHz increments from left to right across columns.

kHz	feet	feet	feet	feet	feet	feet	feet	feet	feet	feet
25500	18.35	18.35	18.35	18.34	18.34	18.34	18.34	18.34	18.34	18.33
25520	18.33	18.33	18.33	18.33	18.33	18.33	18.32	18.32	18.32	18.32
25540	18.32	18.32	18.32	18.31	18.31	18.31	18.31	18.31	18.31	18.31
25560	18.3	18.3	18.3	18.3	18.3	18.3	18.3	18.29	18.29	18.29
25580	18.29	18.29	18.29	18.29	18.28	18.28	18.28	18.28	18.28	18.28
25600	18.28	18.27	18.27	18.27	18.27	18.27	18.27	18.27	18.26	18.26
25620	18.26	18.26	18.26	18.26	18.26	18.25	18.25	18.25	18.25	18.25
25640	18.25	18.25	18.24	18.24	18.24	18.24	18.24	18.24	18.24	18.23
25660	18.23	18.23	18.23	18.23	18.23	18.23	18.22	18.22	18.22	18.22
25680	18.22	18.22	18.22	18.22	18.21	18.21	18.21	18.21	18.21	18.21
25700	18.21	18.2	18.2	18.2	18.2	18.2	18.2	18.2	18.19	18.19
25720	18.19	18.19	18.19	18.19	18.19	18.18	18.18	18.18	18.18	18.18
25740	18.18	18.18	18.17	18.17	18.17	18.17	18.17	18.17	18.17	18.16
25760	18.16	18.16	18.16	18.16	18.16	18.16	18.15	18.15	18.15	18.15
25780	18.15	18.15	18.15	18.14	18.14	18.14	18.14	18.14	18.14	18.14
25800	18.13	18.13	18.13	18.13	18.13	18.13	18.13	18.12	18.12	18.12
25820	18.12	18.12	18.12	18.12	18.11	18.11	18.11	18.11	18.11	18.11
25840	18.11	18.11	18.1	18.1	18.1	18.1	18.1	18.1	18.1	18.09
25860	18.09	18.09	18.09	18.09	18.09	18.09	18.08	18.08	18.08	18.08
25880	18.08	18.08	18.08	18.07	18.07	18.07	18.07	18.07	18.07	18.07
25900	18.06	18.06	18.06	18.06	18.06	18.06	18.06	18.05	18.05	18.05
25920	18.05	18.05	18.05	18.05	18.04	18.04	18.04	18.04	18.04	18.04
25940	18.04	18.04	18.03	18.03	18.03	18.03	18.03	18.03	18.03	18.02
25960	18.02	18.02	18.02	18.02	18.02	18.02	18.01	18.01	18.01	18.01
25980	18.01	18.01	18.01	18	18	18	18	18	18	18
26000	18	17.99	17.99	17.99	17.99	17.99	17.99	17.99	17.98	17.98

TABLE 1-16
HALF-WAVELENGTH DIPOLE
26500 - 27000 KILOHERTZ
USE: INT'L. BROADCAST

Measurements, in feet, advance in 1 kHz increments from left to right across columns.

kHz	feet	feet	feet	feet	feet	feet	feet	feet	feet	feet
26500	17.66	17.65	17.65	17.65	17.65	17.65	17.65	17.65	17.64	17.64
26520	17.64	17.64	17.64	17.64	17.64	17.64	17.63	17.63	17.63	17.63
26540	17.63	17.63	17.63	17.62	17.62	17.62	17.62	17.62	17.62	17.62
26560	17.62	17.61	17.61	17.61	17.61	17.61	17.61	17.61	17.6	17.6
26580	17.6	17.6	17.6	17.6	17.6	17.6	17.59	17.59	17.59	17.59
26600	17.59	17.59	17.59	17.59	17.58	17.58	17.58	17.58	17.58	17.58
26620	17.58	17.57	17.57	17.57	17.57	17.57	17.57	17.57	17.57	17.56
26640	17.56	17.56	17.56	17.56	17.56	17.56	17.55	17.55	17.55	17.55
26660	17.55	17.55	17.55	17.55	17.54	17.54	17.54	17.54	17.54	17.54
26680	17.54	17.53	17.53	17.53	17.53	17.53	17.53	17.53	17.53	17.52
26700	17.52	17.52	17.52	17.52	17.52	17.52	17.52	17.51	17.51	17.51
26720	17.51	17.51	17.51	17.51	17.5	17.5	17.5	17.5	17.5	17.5
26740	17.5	17.5	17.49	17.49	17.49	17.49	17.49	17.49	17.49	17.49
26760	17.48	17.48	17.48	17.48	17.48	17.48	17.48	17.47	17.47	17.47
26780	17.47	17.47	17.47	17.47	17.47	17.46	17.46	17.46	17.46	17.46
26800	17.46	17.46	17.46	17.45	17.45	17.45	17.45	17.45	17.45	17.45
26820	17.44	17.44	17.44	17.44	17.44	17.44	17.44	17.44	17.43	17.43
26840	17.43	17.43	17.43	17.43	17.43	17.43	17.42	17.42	17.42	17.42
26860	17.42	17.42	17.42	17.41	17.41	17.41	17.41	17.41	17.41	17.41
26880	17.41	17.4	17.4	17.4	17.4	17.4	17.4	17.4	17.4	17.39
26900	17.39	17.39	17.39	17.39	17.39	17.39	17.39	17.38	17.38	17.38
26920	17.38	17.38	17.38	17.38	17.37	17.37	17.37	17.37	17.37	17.37
26940	17.37	17.37	17.36	17.36	17.36	17.36	17.36	17.36	17.36	17.36
26960	17.35	17.35	17.35	17.35	17.35	17.35	17.35	17.35	17.34	17.34
26980	17.34	17.34	17.34	17.34	17.34	17.33	17.33	17.33	17.33	17.33
27000	17.33	17.33	17.33	17.32	17.32	17.32	17.32	17.32	17.32	17.32

TABLE 1-17
HALF-WAVELENGTH DIPOLE
28500 - 29000 KILOHERTZ
USE: AMATEUR

Measurements, in feet, advance in 1 kHz increments from left to right across columns.

kHz	feet	feet	feet	feet	feet	feet	feet	feet	feet	feet
28500	16.42	16.41	16.41	16.41	16.41	16.41	16.41	16.41	16.41	16.41
28520	16.4	16.4	16.4	16.4	16.4	16.4	16.4	16.4	16.4	16.39
28540	16.39	16.39	16.39	16.39	16.39	16.39	16.39	16.38	16.38	16.38
28560	16.38	16.38	16.38	16.38	16.38	16.38	16.37	16.37	16.37	16.37
28580	16.37	16.37	16.37	16.37	16.37	16.36	16.36	16.36	16.36	16.36
28600	16.36	16.36	16.36	16.36	16.35	16.35	16.35	16.35	16.35	16.35
28620	16.35	16.35	16.34	16.34	16.34	16.34	16.34	16.34	16.34	16.34
28640	16.34	16.33	16.33	16.33	16.33	16.33	16.33	16.33	16.33	16.33
28660	16.32	16.32	16.32	16.32	16.32	16.32	16.32	16.32	16.32	16.31
28680	16.31	16.31	16.31	16.31	16.31	16.31	16.31	16.31	16.3	16.3
28700	16.3	16.3	16.3	16.3	16.3	16.3	16.29	16.29	16.29	16.29
28720	16.29	16.29	16.29	16.29	16.29	16.28	16.28	16.28	16.28	16.28
28740	16.28	16.28	16.28	16.28	16.27	16.27	16.27	16.27	16.27	16.27
28760	16.27	16.27	16.27	16.26	16.26	16.26	16.26	16.26	16.26	16.26
28780	16.26	16.26	16.25	16.25	16.25	16.25	16.25	16.25	16.25	16.25
28800	16.25	16.24	16.24	16.24	16.24	16.24	16.24	16.24	16.24	16.23
28820	16.23	16.23	16.23	16.23	16.23	16.23	16.23	16.23	16.22	16.22
28840	16.22	16.22	16.22	16.22	16.22	16.22	16.22	16.21	16.21	16.21
28860	16.21	16.21	16.21	16.21	16.21	16.21	16.2	16.2	16.2	16.2
28880	16.2	16.2	16.2	16.2	16.2	16.19	16.19	16.19	16.19	16.19
28900	16.19	16.19	16.19	16.19	16.18	16.18	16.18	16.18	16.18	16.18
28920	16.18	16.18	16.18	16.17	16.17	16.17	16.17	16.17	16.17	16.17
28940	16.17	16.17	16.16	16.16	16.16	16.16	16.16	16.16	16.16	16.16
28960	16.16	16.15	16.15	16.15	16.15	16.15	16.15	16.15	16.15	16.15
28980	16.14	16.14	16.14	16.14	16.14	16.14	16.14	16.14	16.14	16.13
29000	16.13	16.13	16.13	16.13	16.13	16.13	16.13	16.13	16.12	16.12

TABLE 1 - 18
HALF WAVELENGTH DIPOLE
29000 - 29500 KILOHERTZ
USE: AMATEUR

Measurements, in feet, advance in 1 kHz increments from left to right across columns.

kHz	feet	feet	feet	feet	feet	feet	feet	feet	feet	feet
29000	16.13	16.13	16.13	16.13	16.13	16.13	16.13	16.13	16.12	16.12
29020	16.12	16.12	16.12	16.12	16.12	16.12	16.12	16.11	16.11	16.11
29040	16.11	16.11	16.11	16.11	16.11	16.11	16.1	16.1	16.1	16.1
29060	16.1	16.1	16.1	16.1	16.1	16.09	16.09	16.09	16.09	16.09
29080	16.09	16.09	16.09	16.09	16.08	16.08	16.08	16.08	16.08	16.08
29100	16.08	16.08	16.08	16.07	16.07	16.07	16.07	16.07	16.07	16.07
29120	16.07	16.07	16.06	16.06	16.06	16.06	16.06	16.06	16.06	16.06
29140	16.06	16.05	16.05	16.05	16.05	16.05	16.05	16.05	16.05	16.05
29160	16.04	16.04	16.04	16.04	16.04	16.04	16.04	16.04	16.04	16.03
29180	16.03	16.03	16.03	16.03	16.03	16.03	16.03	16.03	16.02	16.02
29200	16.02	16.02	16.02	16.02	16.02	16.02	16.02	16.01	16.01	16.01
29220	16.01	16.01	16.01	16.01	16.01	16.01	16	16	16	16
29240	16	16	16	16	16	16	15.99	15.99	15.99	15.99
29260	15.99	15.99	15.99	15.99	15.99	15.98	15.98	15.98	15.98	15.98
29280	15.98	15.98	15.98	15.98	15.97	15.97	15.97	15.97	15.97	15.97
29300	15.97	15.97	15.97	15.96	15.96	15.96	15.96	15.96	15.96	15.96
29320	15.96	15.96	15.95	15.95	15.95	15.95	15.95	15.95	15.95	15.95
29340	15.95	15.94	15.94	15.94	15.94	15.94	15.94	15.94	15.94	15.94
29360	15.94	15.93	15.93	15.93	15.93	15.93	15.93	15.93	15.93	15.93
29380	15.92	15.92	15.92	15.92	15.92	15.92	15.92	15.92	15.92	15.91
29400	15.91	15.91	15.91	15.91	15.91	15.91	15.91	15.91	15.9	15.9
29420	15.9	15.9	15.9	15.9	15.9	15.9	15.9	15.89	15.89	15.89
29440	15.89	15.89	15.89	15.89	15.89	15.89	15.89	15.88	15.88	15.88
29460	15.88	15.88	15.88	15.88	15.88	15.88	15.87	15.87	15.87	15.87
29480	15.87	15.87	15.87	15.87	15.87	15.86	15.86	15.86	15.86	15.86
29500	15.86	15.86	15.86	15.86	15.86	15.85	15.85	15.85	15.85	15.85

TABLE 1-19
HALF-WAVELENGTH DIPOLE
29500 - 30000 KILOHERTZ
USE: AMATEUR

Measurements, in feet, advance in 1 kHz increments from left to right across columns.

kHz	feet	feet	feet	feet	feet	feet	feet	feet	feet	feet
29500	15.86	15.86	15.86	15.86	15.86	15.85	15.85	15.85	15.85	15.85
29520	15.85	15.85	15.85	15.85	15.84	15.84	15.84	15.84	15.84	15.84
29540	15.84	15.84	15.84	15.83	15.83	15.83	15.83	15.83	15.83	15.83
29560	15.83	15.83	15.83	15.82	15.82	15.82	15.82	15.82	15.82	15.82
29580	15.82	15.82	15.81	15.81	15.81	15.81	15.81	15.81	15.81	15.81
29600	15.81	15.8	15.8	15.8	15.8	15.8	15.8	15.8	15.8	15.8
29620	15.8	15.79	15.79	15.79	15.79	15.79	15.79	15.79	15.79	15.79
29640	15.78	15.78	15.78	15.78	15.78	15.78	15.78	15.78	15.78	15.77
29660	15.77	15.77	15.77	15.77	15.77	15.77	15.77	15.77	15.77	15.76
29680	15.76	15.76	15.76	15.76	15.76	15.76	15.76	15.76	15.75	15.75
29700	15.75	15.75	15.75	15.75	15.75	15.75	15.75	15.75	15.74	15.74
29720	15.74	15.74	15.74	15.74	15.74	15.74	15.74	15.73	15.73	15.73
29740	15.73	15.73	15.73	15.73	15.73	15.73	15.73	15.72	15.72	15.72
29760	15.72	15.72	15.72	15.72	15.72	15.72	15.71	15.71	15.71	15.71
29780	15.71	15.71	15.71	15.71	15.71	15.7	15.7	15.7	15.7	15.7
29800	15.7	15.7	15.7	15.7	15.7	15.69	15.69	15.69	15.69	15.69
29820	15.69	15.69	15.69	15.69	15.68	15.68	15.68	15.68	15.68	15.68
29840	15.68	15.68	15.68	15.68	15.67	15.67	15.67	15.67	15.67	15.67
29860	15.67	15.67	15.67	15.66	15.66	15.66	15.66	15.66	15.66	15.66
29880	15.66	15.66	15.66	15.65	15.65	15.65	15.65	15.65	15.65	15.65
29900	15.65	15.65	15.65	15.64	15.64	15.64	15.64	15.64	15.64	15.64
29920	15.64	15.64	15.63	15.63	15.63	15.63	15.63	15.63	15.63	15.63
29940	15.63	15.63	15.62	15.62	15.62	15.62	15.62	15.62	15.62	15.62
29960	15.62	15.61	15.61	15.61	15.61	15.61	15.61	15.61	15.61	15.61
29980	15.61	15.6	15.6	15.6	15.6	15.6	15.6	15.6	15.6	15.6
30000	15.6	15.59	15.59	15.59	15.59	15.59	15.59	15.59	15.59	15.59

Chapter 2
Inverted-Vee Antennas

An inverted-vee antenna is a special type of wavelength dipole in which the feedpoint (the center) is raised above the earth's surface, and the ends droop down toward the earth. An example of the inverted-vee is shown in Fig. 2-1.

The angle between the radiators should be kept between 90 degrees and 120 degrees. Note that this angle will cause both the feedpoint impedance and the required length of the quarter-wavelength sections to vary from the values given for the dipole. In most cases, the lengths of the sections will be from 1 percent to six percent longer in the inverted-vee than in the straight dipole. An SWR meter is usually required to trim the lengths of the sections to optimum.

The gain of the inverted-vee is lower than that of the dipole, because the pattern is less directive than that of the regular dipole. But in return for this, we obtain a lower angle of radiation, generally considered good for DX work. The dipole, by the way, is the reference for most antenna gain measurements.

Inverted-vee antennas are popular on the lower amateur radio bands (i.e. 75 through 20 meters) in locations where there is too little horizontal length for a regular dipole. I have used an inverted-vee antenna at my location. The apex of the antenna was mounted to a wooden pole at the peak of my house roof, and the ends were fastened to insulated supports near the ground.

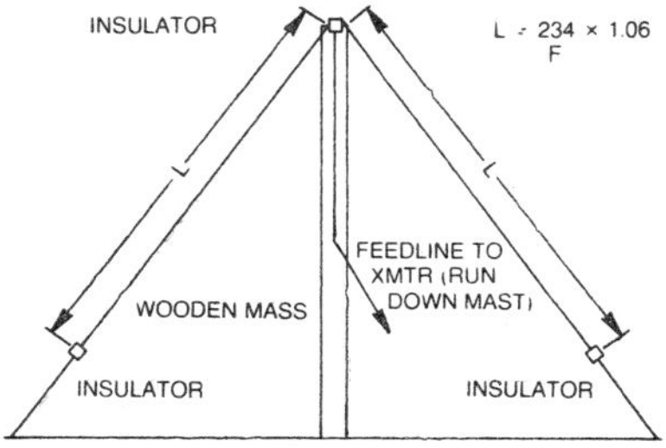

Fig. 2-1. Inverted-vee dipole construction.

TABLE 2-1
HALF WAVELENGTH INVERTED-VEE
300 - 500 KILOHERTZ
USE: MARINE/RADIO NAVIGATION

Measurements, in feet, advance in 1 kHz increments from left to right across columns.

kHz	feet	feet	feet	feet	feet	feet	feet	feet	feet	feet
300	1653.33	1647.84	1642.38	1636.96	1631.57	1626.22	1620.91	1615.63	1610.38	1605.17
310	1600	1594.85	1589.74	1584.66	1579.61	1574.6	1569.62	1564.66	1559.74	1554.85
320	1550	1545.17	1540.37	1535.6	1530.86	1526.15	1521.47	1516.81	1512.19	1507.59
330	1503.03	1498.48	1493.97	1489.48	1485.02	1480.59	1476.19	1471.81	1467.45	1463.12
340	1458.82	1454.54	1450.29	1446.06	1441.86	1437.68	1433.52	1429.39	1425.28	1421.2
350	1417.14	1413.1	1409.09	1405.09	1401.12	1397.18	1393.25	1389.35	1385.47	1381.61
360	1377.77	1373.96	1370.16	1366.39	1362.63	1358.9	1355.19	1351.49	1347.82	1344.17
370	1340.54	1336.92	1333.33	1329.75	1326.2	1322.66	1319.14	1315.64	1312.16	1308.7
380	1305.26	1301.83	1298.42	1295.03	1291.66	1288.31	1284.97	1281.65	1278.35	1275.06
390	1271.79	1268.54	1265.3	1262.08	1258.88	1255.69	1252.52	1249.37	1246.23	1243.1
400	1240	1236.9	1233.83	1230.76	1227.72	1224.69	1221.67	1218.67	1215.68	1212.71
410	1209.75	1206.81	1203.88	1200.96	1198.06	1195.18	1192.3	1189.44	1186.6	1183.77
420	1180.95	1178.14	1175.35	1172.57	1169.81	1167.05	1164.31	1161.59	1158.87	1156.17
430	1153.48	1150.81	1148.14	1145.49	1142.85	1140.22	1137.61	1135.01	1132.42	1129.84
440	1127.27	1124.71	1122.17	1119.63	1117.11	1114.6	1112.1	1109.61	1107.14	1104.67
450	1102.22	1099.77	1097.34	1094.92	1092.51	1090.1	1087.71	1085.33	1082.96	1080.61
460	1078.26	1075.92	1073.59	1071.27	1068.96	1066.66	1064.37	1062.09	1059.82	1057.56
470	1055.31	1053.07	1050.84	1048.62	1046.41	1044.21	1042.01	1039.83	1037.65	1035.49
480	1033.33	1031.18	1029.04	1026.91	1024.79	1022.68	1020.57	1018.48	1016.39	1014.31
490	1012.24	1010.18	1008.13	1006.08	1004.04	1002.02	1000	997.98	995.98	993.98
500	992	990.01	988.04	986.08	984.12	982.17	980.23	978.3	976.37	974.45

TABLE2-2
HALF WAVELENGTH INVERTED-VEE DIPOLE
540 - 1605 KILOHERTZ
USE: AM BROADCASTING

Measurements, in feet, advance in 1 kHz increments from left to right across columns.

kHz	feet	feet	feet	feet	feet	feet	feet	feet	feet	feet
540	918.66	910.23	901.96	893.83	885.85	878.01	870.31	862.74	855.31	848
590	840.81	833.74	826.8	819.96	813.24	806.63	800.12	793.72	787.42	781.22
640	775.12	769.11	763.2	757.37	751.63	745.98	740.41	734.93	729.52	724.2
690	718.95	713.78	708.68	703.65	698.7	693.81	689	684.24	679.56	674.93
740	670.37	665.87	661.44	657.05	652.73	648.47	644.25	640.1	636	631.94
790	627.94	624	620.09	616.24	612.44	608.68	604.97	601.3	597.68	594.1
840	590.57	587.07	583.62	580.21	576.83	573.5	570.2	566.94	563.72	560.54
890	557.39	554.27	551.2	548.15	545.14	542.16	539.21	536.3	533.41	530.56
940	527.74	524.95	522.18	519.45	516.75	514.07	511.42	508.8	506.2	503.63
990	501.09	498.57	496.08	493.61	491.16	488.74	486.35	483.98	481.63	479.3
1040	477	474.71	472.45	470.21	468	465.8	463.62	461.46	459.33	457.21
1090	455.11	453.04	450.98	448.94	446.91	444.91	442.92	440.96	439	437.07
1140	435.15	433.25	431.37	429.5	427.65	425.81	424	422.19	420.4	418.63
1190	416.87	415.12	413.4	411.68	409.98	408.29	406.62	404.96	403.31	401.68
1240	400.06	398.45	396.86	395.28	393.71	392.15	390.61	389.08	387.56	386.05
1290	384.55	383.07	381.6	380.13	378.68	377.24	375.81	374.4	372.99	371.59
1340	370.2	368.83	367.46	366.11	364.76	363.42	362.1	360.78	359.47	358.18
1390	356.89	355.61	354.34	353.08	351.82	350.58	349.35	348.12	346.9	345.7
1440	344.5	343.3	342.12	340.94	339.78	338.62	337.46	336.32	335.18	334.06
1490	332.93	331.82	330.72	329.62	328.52	327.44	326.36	325.29	324.23	323.17
1540	322.12	321.08	320.05	319.02	318	316.98	315.97	314.97	313.97	312.98
1590	312	311.02	310.04	309.08	308.12	307.17	306.22	305.28	304.34	303.41

TABLE2-3
HALF WAVELENGTH INVERTED-VEE DIPOLE
1800 - 2000 KILOHERTZ
USE: AMATEUR

Measurements, in feet, advance in 1 kHz increments from left to right across columns.

kHz	feet	feet	feet	feet	feet	feet	feet	feet	feet	feet
1800	275.6	275.44	275.29	275.14	274.98	274.83	274.68	274.53	274.38	274.22
1810	274.07	273.92	273.77	273.62	273.47	273.32	273.17	273.02	272.87	272.72
1820	272.57	272.42	272.27	272.12	271.97	271.82	271.67	271.52	271.37	271.23
1830	271.08	270.93	270.78	270.63	270.49	270.34	270.19	270.04	269.9	269.75
1840	269.6	269.46	269.31	269.16	269.02	268.87	268.73	268.58	268.44	268.29
1850	268.15	268	267.86	267.71	267.57	267.42	267.28	267.14	266.99	266.85
1860	266.7	266.56	266.42	266.28	266.13	265.99	265.85	265.7	265.56	265.42
1870	265.28	265.14	265	264.85	264.71	264.57	264.43	264.29	264.15	264.01
1880	263.87	263.73	263.59	263.45	263.31	263.17	263.03	262.89	262.75	262.61
1890	262.47	262.33	262.19	262.06	261.92	261.78	261.64	261.5	261.36	261.23
1900	261.09	260.95	260.82	260.68	260.54	260.4	260.27	260.13	260	259.86
1910	259.72	259.59	259.45	259.32	259.18	259.04	258.91	258.77	258.64	258.5
1920	258.37	258.24	258.1	257.97	257.83	257.7	257.57	257.43	257.3	257.16
1930	257.03	256.9	256.77	256.63	256.5	256.37	256.23	256.1	255.97	255.84
1940	255.71	255.57	255.44	255.31	255.18	255.05	254.92	254.79	254.66	254.53
1950	254.4	254.26	254.13	254	253.87	253.74	253.61	253.49	253.36	253.23
1960	253.1	252.97	252.84	252.71	252.58	252.45	252.32	252.2	252.07	251.94
1970	251.81	251.68	251.56	251.43	251.3	251.17	251.05	250.92	250.79	250.67
1980	250.54	250.41	250.29	250.16	250.04	249.91	249.78	249.66	249.53	249.41
1990	249.28	249.16	249.03	248.91	248.78	248.66	248.53	248.41	248.28	248.16
2000	248.04	247.91	247.79	247.66	247.54	247.42	247.29	247.17	247.05	246.92

TABLE 2-4
HALF WAVELENGTH INVERTED-VEE DIPOLE
3500 - 4000 KILOHERTZ
USE: AMATEUR

Measurements, in feet, advance in 1 kHz increments from left to right across columns.

kHz	feet	feet	feet	feet	feet	feet	feet	feet	feet	feet
3500	141.73	141.65	141.57	141.49	141.41	141.33	141.25	141.17	141.09	141.01
3520	140.93	140.85	140.77	140.69	140.61	140.53	140.45	140.37	140.29	140.21
3540	140.13	140.05	139.97	139.89	139.81	139.74	139.66	139.58	139.5	139.42
3560	139.34	139.27	139.19	139.11	139.03	138.95	138.88	138.8	138.72	138.64
3580	138.56	138.49	138.41	138.33	138.26	138.18	138.1	138.03	137.95	137.87
3600	137.8	137.72	137.64	137.57	137.49	137.41	137.34	137.26	137.19	137.11
3620	137.03	136.96	136.88	136.81	136.73	136.66	136.58	136.51	136.43	136.36
3640	136.28	136.21	136.13	136.06	135.98	135.91	135.83	135.76	135.68	135.61
3660	135.54	135.46	135.39	135.31	135.24	135.17	135.09	135.02	134.95	134.87
3680	134.8	134.73	134.65	134.58	134.51	134.43	134.36	134.29	134.22	134.14
3700	134.07	134	133.93	133.85	133.78	133.71	133.64	133.57	133.49	133.42
3720	133.35	133.28	133.21	133.14	133.06	132.99	132.92	132.85	132.78	132.71
3740	132.64	132.57	132.5	132.42	132.35	132.28	132.21	132.14	132.07	132
3760	131.93	131.86	131.79	131.72	131.65	131.58	131.51	131.44	131.37	131.3
3780	131.23	131.16	131.09	131.03	130.96	130.89	130.82	130.75	130.68	130.61
3800	130.54	130.47	130.41	130.34	130.27	130.2	130.13	130.06	130	129.93
3820	129.86	129.79	129.72	129.66	129.59	129.52	129.45	129.38	129.32	129.25
3840	129.18	129.12	129.05	128.98	128.91	128.85	128.78	128.71	128.65	128.58
3860	128.51	128.45	128.38	128.31	128.25	128.18	128.11	128.05	127.98	127.92
3880	127.85	127.78	127.72	127.65	127.59	127.52	127.46	127.39	127.33	127.26
3900	127.2	127.13	127.06	127	126.93	126.87	126.8	126.74	126.68	126.61
3920	126.55	126.48	126.42	126.35	126.29	126.22	126.16	126.1	126.03	125.97
3940	125.9	125.84	125.78	125.71	125.65	125.58	125.52	125.46	125.39	125.33
3960	125.27	125.2	125.14	125.08	125.02	124.95	124.89	124.83	124.76	124.7
3980	124.64	124.58	124.51	124.45	124.39	124.33	124.26	124.2	124.14	124.08
4000	124.02	123.95	123.89	123.83	123.77	123.71	123.64	123.58	123.52	123.46

TABLE 2-5
HALF WAVELENGTH INVERTED-VEE DIPOLE
5500 - 6000 KILOHERTZ
USE: INT'L. BROADCAST

Measurements, in feet, advance in 1 kHz increments from left to right across columns.

kHz	feet	feet	feet	feet	feet	feet	feet	feet	feet	feet
5500	90.19	90.16	90.13	90.09	90.06	90.03	90	89.96	89.93	89.9
5520	89.86	89.83	89.8	89.77	89.73	89.7	89.67	89.64	89.6	89.57
5540	89.54	89.51	89.48	89.44	89.41	89.38	89.35	89.31	89.28	89.25
5560	89.22	89.19	89.15	89.12	89.09	89.06	89.03	88.99	88.96	88.93
5580	88.9	88.87	88.83	88.8	88.77	88.74	88.71	88.68	88.64	88.61
5600	88.58	88.55	88.52	88.49	88.45	88.42	88.39	88.36	88.33	88.3
5620	88.27	88.23	88.2	88.17	88.14	88.11	88.08	88.05	88.01	87.98
5640	87.95	87.92	87.89	87.86	87.83	87.8	87.77	87.73	87.7	87.67
5660	87.64	87.61	87.58	87.55	87.52	87.49	87.46	87.43	87.39	87.36
5680	87.33	87.3	87.27	87.24	87.21	87.18	87.15	87.12	87.09	87.06
5700	87.03	87	86.97	86.94	86.9	86.87	86.84	86.81	86.78	86.75
5720	86.72	86.69	86.66	86.63	86.6	86.57	86.54	86.51	86.48	86.45
5740	86.42	86.39	86.36	86.33	86.3	86.27	86.24	86.21	86.18	86.15
5760	86.12	86.09	86.06	86.03	86	85.97	85.94	85.91	85.88	85.85
5780	85.82	85.79	85.76	85.73	85.7	85.67	85.64	85.61	85.59	85.56
5800	85.53	85.5	85.47	85.44	85.41	85.38	85.35	85.32	85.29	85.26
5820	85.23	85.2	85.17	85.14	85.12	85.09	85.06	85.03	85	84.97
5840	84.94	84.91	84.88	84.85	84.82	84.8	84.77	84.74	84.71	84.68
5860	84.65	84.62	84.59	84.56	84.53	84.51	84.48	84.45	84.42	84.39
5880	84.36	84.33	84.3	84.28	84.25	84.22	84.19	84.16	84.13	84.1
5900	84.08	84.05	84.02	83.99	83.96	83.93	83.91	83.88	83.85	83.82
5920	83.79	83.76	83.74	83.71	83.68	83.65	83.62	83.59	83.57	83.54
5940	83.51	83.48	83.45	83.43	83.4	83.37	83.34	83.31	83.29	83.26
5960	83.23	83.2	83.17	83.15	83.12	83.09	83.06	83.03	83.01	82.98
5980	82.95	82.92	82.9	82.87	82.84	82.81	82.79	82.76	82.73	82.7
6000	82.68	82.65	82.62	82.59	82.56	82.54	82.51	82.48	82.46	82.43

TABLE 2-6
HALF WAVELENGTH INVERTED-VEE DIPOLE
6000 - 6500 KILOHERTZ
USE: INT´L. BROADCAST

Measurements, in feet, advance in 1 kHz increments from left to right across columns.

kHz	feet	feet	feet	feet	feet	feet	feet	feet	feet	feet
6000	82.68	82.65	82.62	82.59	82.56	82.54	82.51	82.48	82.46	82.43
6020	82.4	82.37	82.35	82.32	82.29	82.26	82.24	82.21	82.18	82.15
6040	82.13	82.1	82.07	82.05	82.02	81.99	81.96	81.94	81.91	81.88
6060	81.86	81.83	81.8	81.78	81.75	81.72	81.69	81.67	81.64	81.61
6080	81.59	81.56	81.53	81.51	81.48	81.45	81.43	81.4	81.37	81.35
6100	81.32	81.29	81.27	81.24	81.21	81.19	81.16	81.13	81.11	81.08
6120	81.05	81.03	81	80.97	80.95	80.92	80.9	80.87	80.84	80.82
6140	80.79	80.76	80.74	80.71	80.68	80.66	80.63	80.61	80.58	80.55
6160	80.53	80.5	80.48	80.45	80.42	80.4	80.37	80.34	80.32	80.29
6180	80.27	80.24	80.21	80.19	80.16	80.14	80.11	80.09	80.06	80.03
6200	80.01	79.98	79.96	79.93	79.9	79.88	79.85	79.83	79.8	79.78
6220	79.75	79.72	79.7	79.67	79.65	79.62	79.6	79.57	79.55	79.52
6240	79.5	79.47	79.44	79.42	79.39	79.37	79.34	79.32	79.29	79.27
6260	79.24	79.22	79.19	79.17	79.14	79.11	79.09	79.06	79.04	79.01
6280	78.99	78.96	78.94	78.91	78.89	78.86	78.84	78.81	78.79	78.76
6300	78.74	78.71	78.69	78.66	78.64	78.61	78.59	78.56	78.54	78.51
6320	78.49	78.46	78.44	78.41	78.39	78.36	78.34	78.32	78.29	78.27
6340	78.24	78.22	78.19	78.17	78.14	78.12	78.09	78.07	78.04	78.02
6360	78	77.97	77.95	77.92	77.9	77.87	77.85	77.82	77.8	77.77
6380	77.75	77.73	77.7	77.68	77.65	77.63	77.6	77.58	77.56	77.53
6400	77.51	77.48	77.46	77.43	77.41	77.39	77.36	77.34	77.31	77.29
6420	77.27	77.24	77.22	77.19	77.17	77.15	77.12	77.1	77.07	77.05
6440	77.03	77	76.98	76.95	76.93	76.91	76.88	76.86	76.84	76.81
6460	76.79	76.76	76.74	76.72	76.69	76.67	76.65	76.62	76.6	76.57
6480	76.55	76.53	76.5	76.48	76.46	76.43	76.41	76.39	76.36	76.34
6500	76.32	76.29	76.27	76.24	76.22	76.2	76.17	76.15	76.13	76.1

TABLE 2-7
HALF WAVELENGTH INVERTED-VEE DIPOLE
7000 - 7300 KILOHERTZ
USE: AMATEUR

Measurements, in feet, advance in 1 kHz increments from left to right across columns.

kHz	feet	feet	feet	feet	feet	feet	feet	feet	feet	feet
7000	70.86	70.84	70.82	70.8	70.78	70.76	70.74	70.72	70.7	70.68
7020	70.66	70.64	70.62	70.6	70.58	70.56	70.54	70.52	70.5	70.48
7040	70.46	70.44	70.42	70.4	70.38	70.36	70.34	70.32	70.3	70.28
7060	70.26	70.24	70.22	70.2	70.18	70.16	70.14	70.12	70.1	70.08
7080	70.06	70.04	70.02	70	69.98	69.96	69.94	69.92	69.9	69.89
7100	69.87	69.85	69.83	69.81	69.79	69.77	69.75	69.73	69.71	69.69
7120	69.67	69.65	69.63	69.61	69.59	69.57	69.55	69.53	69.51	69.49
7140	69.47	69.45	69.44	69.42	69.4	69.38	69.36	69.34	69.32	69.3
7160	69.28	69.26	69.24	69.22	69.2	69.18	69.16	69.14	69.13	69.11
7180	69.09	69.07	69.05	69.03	69.01	68.99	68.97	68.95	68.93	68.91
7200	68.9	68.88	68.86	68.84	68.82	68.8	68.78	68.76	68.74	68.72
7220	68.7	68.69	68.67	68.65	68.63	68.61	68.59	68.57	68.55	68.53
7240	68.51	68.5	68.48	68.46	68.44	68.42	68.4	68.38	68.36	68.34
7260	68.33	68.31	68.29	68.27	68.25	68.23	68.21	68.19	68.18	68.16
7280	68.14	68.12	68.1	68.08	68.06	68.04	68.03	68.01	67.99	67.97
7300	67.95	67.93	67.91	67.9	67.88	67.86	67.84	67.82	67.8	67.78

TABLE 2-8
HALF-WAVELENGTH INVERTED-VEE DIPOLE
9500 - 10000 KILOHERTZ
USE: INT'L. BROADCAST

Measurements, in feet, advance in 1 kHz increments from left to right across columns.

kHz	feet	feet	feet	feet	feet	feet	feet	feet	feet	feet
9500	52.21	52.2	52.19	52.18	52.17	52.16	52.15	52.14	52.13	52.12
9520	52.1	52.09	52.08	52.07	52.06	52.05	52.04	52.03	52.02	52.01
9540	52	51.98	51.97	51.96	51.95	51.94	51.93	51.92	51.91	51.9
9560	51.89	51.88	51.86	51.85	51.84	51.83	51.82	51.81	51.8	51.79
9580	51.78	51.77	51.76	51.75	51.73	51.72	51.71	51.7	51.69	51.68
9600	51.67	51.66	51.65	51.64	51.63	51.62	51.61	51.59	51.58	51.57
9620	51.56	51.55	51.54	51.53	51.52	51.51	51.5	51.49	51.48	51.47
9640	51.46	51.44	51.43	51.42	51.41	51.4	51.39	51.38	51.37	51.36
9660	51.35	51.34	51.33	51.32	51.31	51.3	51.29	51.27	51.26	51.25
9680	51.24	51.23	51.22	51.21	51.2	51.19	51.18	51.17	51.16	51.15
9700	51.14	51.13	51.12	51.11	51.1	51.08	51.07	51.06	51.05	51.04
9720	51.03	51.02	51.01	51	50.99	50.98	50.97	50.96	50.95	50.94
9740	50.93	50.92	50.91	50.9	50.89	50.88	50.86	50.85	50.84	50.83
9760	50.82	50.81	50.8	50.79	50.78	50.77	50.76	50.75	50.74	50.73
9780	50.72	50.71	50.7	50.69	50.68	50.67	50.66	50.65	50.64	50.63
9800	50.62	50.61	50.59	50.58	50.57	50.56	50.55	50.54	50.53	50.52
9820	50.51	50.5	50.49	50.48	50.47	50.46	50.45	50.44	50.43	50.42
9840	50.41	50.4	50.39	50.38	50.37	50.36	50.35	50.34	50.33	50.32
9860	50.31	50.3	50.29	50.28	50.27	50.26	50.25	50.24	50.23	50.22
9880	50.21	50.2	50.19	50.18	50.16	50.15	50.14	50.13	50.12	50.11
9900	50.1	50.09	50.08	50.07	50.06	50.05	50.04	50.03	50.02	50.01
9920	50	49.99	49.98	49.97	49.96	49.95	49.94	49.93	49.92	49.91
9940	49.9	49.89	49.88	49.87	49.86	49.85	49.84	49.83	49.82	49.81
9960	49.8	49.79	49.78	49.77	49.76	49.75	49.74	49.73	49.72	49.71
9980	49.7	49.69	49.68	49.67	49.66	49.65	49.64	49.63	49.62	49.61
10000	49.6	49.59	49.58	49.57	49.56	49.55	49.54	49.53	49.52	49.51

TABLE 2-9
HALF-WAVELENGTH INVERTED-VEE DIPOLE
14000 - 14450 KILOHERTZ
USE: AMATEUR

Measurements, in feet, advance in 1 kHz increments from left to right across columns.

kHz	feet	feet	feet	feet	feet	feet	feet	feet	feet	feet
14000	35.43	35.42	35.42	35.41	35.41	35.4	35.4	35.39	35.39	35.38
14020	35.38	35.37	35.37	35.36	35.36	35.35	35.35	35.34	35.34	35.33
14040	35.33	35.32	35.32	35.31	35.31	35.3	35.3	35.29	35.29	35.28
14060	35.28	35.27	35.27	35.26	35.26	35.25	35.25	35.24	35.24	35.23
14080	35.23	35.22	35.22	35.21	35.21	35.2	35.2	35.19	35.19	35.18
14100	35.18	35.17	35.17	35.16	35.16	35.15	35.15	35.14	35.14	35.13
14120	35.13	35.12	35.12	35.11	35.11	35.1	35.1	35.09	35.09	35.08
14140	35.08	35.07	35.07	35.06	35.06	35.05	35.05	35.04	35.04	35.03
14160	35.03	35.02	35.02	35.01	35.01	35	35	34.99	34.99	34.98
14180	34.98	34.97	34.97	34.96	34.96	34.95	34.95	34.94	34.94	34.94
14200	34.93	34.93	34.92	34.92	34.91	34.91	34.9	34.9	34.89	34.89
14220	34.88	34.88	34.87	34.87	34.86	34.86	34.85	34.85	34.84	34.84
14240	34.83	34.83	34.82	34.82	34.81	34.81	34.8	34.8	34.79	34.79
14260	34.78	34.78	34.77	34.77	34.76	34.76	34.75	34.75	34.74	34.74
14280	34.73	34.73	34.72	34.72	34.72	34.71	34.71	34.7	34.7	34.69
14300	34.69	34.68	34.68	34.67	34.67	34.66	34.66	34.65	34.65	34.64
14320	34.64	34.63	34.63	34.62	34.62	34.61	34.61	34.6	34.6	34.59
14340	34.59	34.58	34.58	34.57	34.57	34.57	34.56	34.56	34.55	34.55
14360	34.54	34.54	34.53	34.53	34.52	34.52	34.51	34.51	34.5	34.5
14380	34.49	34.49	34.48	34.48	34.47	34.47	34.46	34.46	34.45	34.45
14400	34.45	34.44	34.44	34.43	34.43	34.42	34.42	34.41	34.41	34.4
14420	34.4	34.39	34.39	34.38	34.38	34.37	34.37	34.36	34.36	34.35
14440	34.35	34.34	34.34	34.34	34.33	34.33	34.32	34.32	34.31	34.31

TABLE 2-11
HALF WAVELENGTH INVERTED-VEE DIPOLE
15500 - 16000 KILOHERTZ
USE: INT´L. BROADCAST

Measurements, in feet, advance in 1 kHz increments from left to right across columns.

kHz	feet	feet	feet	feet	feet	feet	feet	feet	feet	feet
15500	32	32	31.99	31.99	31.98	31.98	31.98	31.97	31.97	31.96
15520	31.96	31.95	31.95	31.95	31.94	31.94	31.93	31.93	31.93	31.92
15540	31.92	31.91	31.91	31.91	31.9	31.9	31.89	31.89	31.88	31.88
15560	31.88	31.87	31.87	31.86	31.86	31.86	31.85	31.85	31.84	31.84
15580	31.84	31.83	31.83	31.82	31.82	31.82	31.81	31.81	31.8	31.8
15600	31.8	31.79	31.79	31.78	31.78	31.77	31.77	31.77	31.76	31.76
15620	31.75	31.75	31.75	31.74	31.74	31.73	31.73	31.73	31.72	31.72
15640	31.71	31.71	31.71	31.7	31.7	31.69	31.69	31.69	31.68	31.68
15660	31.67	31.67	31.67	31.66	31.66	31.65	31.65	31.64	31.64	31.64
15680	31.63	31.63	31.62	31.62	31.62	31.61	31.61	31.6	31.6	31.6
15700	31.59	31.59	31.58	31.58	31.58	31.57	31.57	31.56	31.56	31.56
15720	31.55	31.55	31.54	31.54	31.54	31.53	31.53	31.52	31.52	31.52
15740	31.51	31.51	31.5	31.5	31.5	31.49	31.49	31.48	31.48	31.48
15760	31.47	31.47	31.46	31.46	31.46	31.45	31.45	31.44	31.44	31.44
15780	31.43	31.43	31.42	31.42	31.42	31.41	31.41	31.4	31.4	31.4
15800	31.39	31.39	31.38	31.38	31.38	31.37	31.37	31.36	31.36	31.36
15820	31.35	31.35	31.34	31.34	31.34	31.33	31.33	31.33	31.32	31.32
15840	31.31	31.31	31.31	31.3	31.3	31.29	31.29	31.29	31.28	31.28
15860	31.27	31.27	31.27	31.26	31.26	31.25	31.25	31.25	31.24	31.24
15880	31.23	31.23	31.23	31.22	31.22	31.21	31.21	31.21	31.2	31.2
15900	31.2	31.19	31.19	31.18	31.18	31.18	31.17	31.17	31.16	31.16
15920	31.16	31.15	31.15	31.14	31.14	31.14	31.13	31.13	31.12	31.12
15940	31.12	31.11	31.11	31.1	31.1	31.1	31.09	31.09	31.09	31.08
15960	31.08	31.07	31.07	31.07	31.06	31.06	31.05	31.05	31.05	31.04
15980	31.04	31.03	31.03	31.03	31.02	31.02	31.02	31.01	31.01	31
16000	31	31	30.99	30.99	30.98	30.98	30.98	30.97	30.97	30.97

TABLE 2-12
HALF WAVELENGTH INVERTED-VEE DIPOLE
16500 - 17000 KILOHERTZ
USE: INT´L. BROADCAST

Measurements, in feet, advance in 1 kHz increments from left to right across columns.

kHz	feet	feet	feet	feet	feet	feet	feet	feet	feet	
16500	30.06	30.06	30.05	30.05	30.05	30.04	30.04	30.03	30.03	30.03
16520	30.02	30.02	30.02	30.01	30.01	30.01	30	30	30	29.99
16540	29.99	29.98	29.98	29.98	29.97	29.97	29.97	29.96	29.96	29.96
16560	29.95	29.95	29.94	29.94	29.94	29.93	29.93	29.93	29.92	29.92
16580	29.92	29.91	29.91	29.9	29.9	29.9	29.89	29.89	29.89	29.88
16600	29.88	29.88	29.87	29.87	29.86	29.86	29.86	29.85	29.85	29.85
16620	29.84	29.84	29.84	29.83	29.83	29.83	29.82	29.82	29.81	29.81
16640	29.81	29.8	29.8	29.8	29.79	29.79	29.79	29.78	29.78	29.78
16660	29.77	29.77	29.76	29.76	29.76	29.75	29.75	29.75	29.74	29.74
16680	29.74	29.73	29.73	29.73	29.72	29.72	29.71	29.71	29.71	29.7
16700	29.7	29.7	29.69	29.69	29.69	29.68	29.68	29.68	29.67	29.67
16720	29.66	29.66	29.66	29.65	29.65	29.65	29.64	29.64	29.64	29.63
16740	29.63	29.63	29.62	29.62	29.62	29.61	29.61	29.6	29.6	29.6
16760	29.59	29.59	29.59	29.58	29.58	29.58	29.57	29.57	29.57	29.56
16780	29.56	29.56	29.55	29.55	29.54	29.54	29.54	29.53	29.53	29.53
16800	29.52	29.52	29.52	29.51	29.51	29.51	29.5	29.5	29.5	29.49
16820	29.49	29.48	29.48	29.48	29.47	29.47	29.47	29.46	29.46	29.46
16840	29.45	29.45	29.45	29.44	29.44	29.44	29.43	29.43	29.43	29.42
16860	29.42	29.41	29.41	29.41	29.4	29.4	29.4	29.39	29.39	29.39
16880	29.38	29.38	29.38	29.37	29.37	29.37	29.36	29.36	29.36	29.35
16900	29.35	29.35	29.34	29.34	29.33	29.33	29.33	29.32	29.32	29.32
16920	29.31	29.31	29.31	29.3	29.3	29.3	29.29	29.29	29.29	29.28
16940	29.28	29.28	29.27	29.27	29.27	29.26	29.26	29.26	29.25	29.25
16960	29.25	29.24	29.24	29.23	29.23	29.23	29.22	29.22	29.22	29.21
16980	29.21	29.21	29.2	29.2	29.2	29.19	29.19	29.19	29.18	29.18
17000	29.18	29.17	29.17	29.17	29.16	29.16	29.16	29.15	29.15	29.15

TABLE 2-13
HALF WAVELENGTH INVERTED-VEE DIPOLE
18500 - 19000 KILOHERTZ
USE: INT'L. BROADCAST

Measurements, in feet, advance in 1 kHz increments from left to right across columns.

kHz	feet	feet	feet	feet	feet	feet	feet	feet	feet	feet
18500	26.81	26.81	26.8	26.8	26.8	26.8	26.79	26.79	26.79	26.78
18520	26.78	26.78	26.78	26.77	26.77	26.77	26.76	26.76	26.76	26.76
18540	26.75	26.75	26.75	26.74	26.74	26.74	26.73	26.73	26.73	26.73
18560	26.72	26.72	26.72	26.71	26.71	26.71	26.71	26.7	26.7	26.7
18580	26.69	26.69	26.69	26.69	26.68	26.68	26.68	26.67	26.67	26.67
18600	26.67	26.66	26.66	26.66	26.65	26.65	26.65	26.65	26.64	26.64
18620	26.64	26.63	26.63	26.63	26.63	26.62	26.62	26.62	26.61	26.61
18640	26.61	26.61	26.6	26.6	26.6	26.59	26.59	26.59	26.59	26.58
18660	26.58	26.58	26.57	26.57	26.57	26.57	26.56	26.56	26.56	26.55
18680	26.55	26.55	26.55	26.54	26.54	26.54	26.53	26.53	26.53	26.53
18700	26.52	26.52	26.52	26.51	26.51	26.51	26.51	26.5	26.5	26.5
18720	26.5	26.49	26.49	26.49	26.48	26.48	26.48	26.48	26.47	26.47
18740	26.47	26.46	26.46	26.46	26.46	26.45	26.45	26.45	26.44	26.44
18760	26.44	26.44	26.43	26.43	26.43	26.42	26.42	26.42	26.42	26.41
18780	26.41	26.41	26.4	26.4	26.4	26.4	26.39	26.39	26.39	26.39
18800	26.38	26.38	26.38	26.37	26.37	26.37	26.37	26.36	26.36	26.36
18820	26.35	26.35	26.35	26.35	26.34	26.34	26.34	26.33	26.33	26.33
18840	26.33	26.32	26.32	26.32	26.32	26.31	26.31	26.31	26.3	26.3
18860	26.3	26.3	26.29	26.29	26.29	26.28	26.28	26.28	26.28	26.27
18880	26.27	26.27	26.26	26.26	26.26	26.26	26.25	26.25	26.25	26.25
18900	26.24	26.24	26.24	26.23	26.23	26.23	26.23	26.22	26.22	26.22
18920	26.21	26.21	26.21	26.21	26.2	26.2	26.2	26.2	26.19	26.19
18940	26.19	26.18	26.18	26.18	26.18	26.17	26.17	26.17	26.17	26.16
18960	26.16	26.16	26.15	26.15	26.15	26.15	26.14	26.14	26.14	26.13
18980	26.13	26.13	26.13	26.12	26.12	26.12	26.12	26.11	26.11	26.11
19000	26.1	26.1	26.1	26.1	26.09	26.09	26.09	26.09	26.08	26.08

TABLE 2-14
HALF WAVELENGTH INVERTED-VEE DIPOLE
21000 - 21450 KILOHERTZ
USE: AMATEUR

Measurements, in feet, advance in 1 kHz increments from left to right across columns.

kHz	feet	feet	feet	feet	feet	feet	feet	feet	feet	feet
21000	23.62	23.62	23.61	23.61	23.61	23.61	23.6	23.6	23.6	23.6
21020	23.6	23.59	23.59	23.59	23.59	23.58	23.58	23.58	23.58	23.58
21040	23.57	23.57	23.57	23.57	23.56	23.56	23.56	23.56	23.56	23.55
21060	23.55	23.55	23.55	23.54	23.54	23.54	23.54	23.53	23.53	23.53
21080	23.53	23.53	23.52	23.52	23.52	23.52	23.51	23.51	23.51	23.51
21100	23.51	23.5	23.5	23.5	23.5	23.49	23.49	23.49	23.49	23.49
21120	23.48	23.48	23.48	23.48	23.47	23.47	23.47	23.47	23.47	23.46
21140	23.46	23.46	23.46	23.45	23.45	23.45	23.45	23.45	23.44	23.44
21160	23.44	23.44	23.43	23.43	23.43	23.43	23.43	23.42	23.42	23.42
21180	23.42	23.41	23.41	23.41	23.41	23.41	23.4	23.4	23.4	23.4
21200	23.4	23.39	23.39	23.39	23.39	23.38	23.38	23.38	23.38	23.38
21220	23.37	23.37	23.37	23.37	23.36	23.36	23.36	23.36	23.36	23.35
21240	23.35	23.35	23.35	23.34	23.34	23.34	23.34	23.34	23.33	23.33
21260	23.33	23.33	23.32	23.32	23.32	23.32	23.32	23.31	23.31	23.31
21280	23.31	23.3	23.3	23.3	23.3	23.3	23.29	23.29	23.29	23.29
21300	23.29	23.28	23.28	23.28	23.28	23.27	23.27	23.27	23.27	23.27
21320	23.26	23.26	23.26	23.26	23.25	23.25	23.25	23.25	23.25	23.24
21340	23.24	23.24	23.24	23.23	23.23	23.23	23.23	23.23	23.22	23.22
21360	23.22	23.22	23.22	23.21	23.21	23.21	23.21	23.2	23.2	23.2
21380	23.2	23.2	23.19	23.19	23.19	23.19	23.18	23.18	23.18	23.18
21400	23.18	23.17	23.17	23.17	23.17	23.17	23.16	23.16	23.16	23.16
21420	23.15	23.15	23.15	23.15	23.15	23.14	23.14	23.14	23.14	23.14
21440	23.13	23.13	23.13	23.13	23.12	23.12	23.12	23.12	23.12	23.11

TABLE 2-15
HALF WAVELENGTH INVERTED-VEE DIPOLE
25500 - 26000 KILOHERTZ
USE: INT´L. BROADCAST

Measurements, in feet, advance in 1 kHz increments from left to right across columns.

kHz	feet	feet	feet	feet	feet	feet	feet	feet	feet	feet
25500	19.45	19.45	19.45	19.44	19.44	19.44	19.44	19.44	19.44	19.44
25520	19.43	19.43	19.43	19.43	19.43	19.43	19.42	19.42	19.42	19.42
25540	19.42	19.42	19.42	19.41	19.41	19.41	19.41	19.41	19.41	19.4
25560	19.4	19.4	19.4	19.4	19.4	19.4	19.39	19.39	19.39	19.39
25580	19.39	19.39	19.39	19.38	19.38	19.38	19.38	19.38	19.38	19.37
25600	19.37	19.37	19.37	19.37	19.37	19.37	19.36	19.36	19.36	19.36
25620	19.36	19.36	19.35	19.35	19.35	19.35	19.35	19.35	19.35	19.34
25640	19.34	19.34	19.34	19.34	19.34	19.34	19.33	19.33	19.33	19.33
25660	19.33	19.33	19.32	19.32	19.32	19.32	19.32	19.32	19.32	19.31
25680	19.31	19.31	19.31	19.31	19.31	19.31	19.3	19.3	19.3	19.3
25700	19.3	19.3	19.29	19.29	19.29	19.29	19.29	19.29	19.29	19.28
25720	19.28	19.28	19.28	19.28	19.28	19.28	19.27	19.27	19.27	19.27
25740	19.27	19.27	19.26	19.26	19.26	19.26	19.26	19.26	19.26	19.25
25760	19.25	19.25	19.25	19.25	19.25	19.25	19.24	19.24	19.24	19.24
25780	19.24	19.24	19.23	19.23	19.23	19.23	19.23	19.23	19.23	19.22
25800	19.22	19.22	19.22	19.22	19.22	19.22	19.21	19.21	19.21	19.21
25820	19.21	19.21	19.21	19.2	19.2	19.2	19.2	19.2	19.2	19.19
25840	19.19	19.19	19.19	19.19	19.19	19.19	19.18	19.18	19.18	19.18
25860	19.18	19.18	19.18	19.17	19.17	19.17	19.17	19.17	19.17	19.16
25880	19.16	19.16	19.16	19.16	19.16	19.16	19.15	19.15	19.15	19.15
25900	19.15	19.15	19.15	19.14	19.14	19.14	19.14	19.14	19.14	19.14
25920	19.13	19.13	19.13	19.13	19.13	19.13	19.13	19.12	19.12	19.12
25940	19.12	19.12	19.12	19.11	19.11	19.11	19.11	19.11	19.11	19.11
25960	19.1	19.1	19.1	19.1	19.1	19.1	19.1	19.09	19.09	19.09
25980	19.09	19.09	19.09	19.09	19.08	19.08	19.08	19.08	19.08	19.08
26000	19.08	19.07	19.07	19.07	19.07	19.07	19.07	19.06	19.06	19.06

TABLE 2-16
HALF WAVELENGTH INVERTED-VEE DIPOLE
26500 - 27000 KILOHERTZ
USE: INT'L. BROADCAST

Measurements, in feet, advance in 1 kHz increments from left to right across columns.

kHz	feet	feet	feet	feet	feet	feet	feet	feet	feet	feet
26500	18.72	18.71	18.71	18.71	18.71	18.71	18.71	18.71	18.7	18.7
26520	18.7	18.7	18.7	18.7	18.7	18.69	18.69	18.69	18.69	18.69
26540	18.69	18.69	18.68	18.68	18.68	18.68	18.68	18.68	18.68	18.67
26560	18.67	18.67	18.67	18.67	18.67	18.67	18.66	18.66	18.66	18.66
26580	18.66	18.66	18.66	18.65	18.65	18.65	18.65	18.65	18.65	18.65
26600	18.64	18.64	18.64	18.64	18.64	18.64	18.64	18.63	18.63	18.63
26620	18.63	18.63	18.63	18.63	18.63	18.62	18.62	18.62	18.62	18.62
26640	18.62	18.62	18.61	18.61	18.61	18.61	18.61	18.61	18.61	18.6
26660	18.6	18.6	18.6	18.6	18.6	18.6	18.59	18.59	18.59	18.59
26680	18.59	18.59	18.59	18.58	18.58	18.58	18.58	18.58	18.58	18.58
26700	18.57	18.57	18.57	18.57	18.57	18.57	18.57	18.57	18.56	18.56
26720	18.56	18.56	18.56	18.56	18.56	18.55	18.55	18.55	18.55	18.55
26740	18.55	18.55	18.54	18.54	18.54	18.54	18.54	18.54	18.54	18.53
26760	18.53	18.53	18.53	18.53	18.53	18.53	18.52	18.52	18.52	18.52
26780	18.52	18.52	18.52	18.52	18.51	18.51	18.51	18.51	18.51	18.51
26800	18.51	18.5	18.5	18.5	18.5	18.5	18.5	18.5	18.49	18.49
26820	18.49	18.49	18.49	18.49	18.49	18.48	18.48	18.48	18.48	18.48
26840	18.48	18.48	18.48	18.47	18.47	18.47	18.47	18.47	18.47	18.47
26860	18.46	18.46	18.46	18.46	18.46	18.46	18.46	18.45	18.45	18.45
26880	18.45	18.45	18.45	18.45	18.44	18.44	18.44	18.44	18.44	18.44
26900	18.44	18.44	18.43	18.43	18.43	18.43	18.43	18.43	18.43	18.42
26920	18.42	18.42	18.42	18.42	18.42	18.42	18.41	18.41	18.41	18.41
26940	18.41	18.41	18.41	18.41	18.4	18.4	18.4	18.4	18.4	18.4
26960	18.4	18.39	18.39	18.39	18.39	18.39	18.39	18.39	18.38	18.38
26980	18.38	18.38	18.38	18.38	18.38	18.38	18.37	18.37	18.37	18.37
27000	18.37	18.37	18.37	18.36	18.36	18.36	18.36	18.36	18.36	18.36

TABLE 2-17
HALF WAVELENGTH INVERTED-VEE DIPOLE
28500 - 29000 KILOHERTZ
USE: AMATEUR

Measurements, in feet, advance in 1 kHz increments from left to right across columns.

kHz	feet	feet	feet	feet	feet	feet	feet	feet	feet	feet
28500	17.4	17.4	17.4	17.4	17.4	17.4	17.39	17.39	17.39	17.39
28520	17.39	17.39	17.39	17.39	17.38	17.38	17.38	17.38	17.38	17.38
28540	17.38	17.38	17.37	17.37	17.37	17.37	17.37	17.37	17.37	17.37
28560	17.36	17.36	17.36	17.36	17.36	17.36	17.36	17.36	17.36	17.35
28580	17.35	17.35	17.35	17.35	17.35	17.35	17.35	17.34	17.34	17.34
28600	17.34	17.34	17.34	17.34	17.34	17.33	17.33	17.33	17.33	17.33
28620	17.33	17.33	17.33	17.32	17.32	17.32	17.32	17.32	17.32	17.32
28640	17.32	17.32	17.31	17.31	17.31	17.31	17.31	17.31	17.31	17.31
28660	17.3	17.3	17.3	17.3	17.3	17.3	17.3	17.3	17.29	17.29
28680	17.29	17.29	17.29	17.29	17.29	17.29	17.28	17.28	17.28	17.28
28700	17.28	17.28	17.28	17.28	17.28	17.27	17.27	17.27	17.27	17.27
28720	17.27	17.27	17.27	17.26	17.26	17.26	17.26	17.26	17.26	17.26
28740	17.26	17.25	17.25	17.25	17.25	17.25	17.25	17.25	17.25	17.25
28760	17.24	17.24	17.24	17.24	17.24	17.24	17.24	17.24	17.23	17.23
28780	17.23	17.23	17.23	17.23	17.23	17.23	17.22	17.22	17.22	17.22
28800	17.22	17.22	17.22	17.22	17.22	17.21	17.21	17.21	17.21	17.21
28820	17.21	17.21	17.21	17.2	17.2	17.2	17.2	17.2	17.2	17.2
28840	17.2	17.19	17.19	17.19	17.19	17.19	17.19	17.19	17.19	17.19
28860	17.18	17.18	17.18	17.18	17.18	17.18	17.18	17.18	17.17	17.17
28880	17.17	17.17	17.17	17.17	17.17	17.17	17.17	17.16	17.16	17.16
28900	17.16	17.16	17.16	17.16	17.16	17.15	17.15	17.15	17.15	17.15
28920	17.15	17.15	17.15	17.14	17.14	17.14	17.14	17.14	17.14	17.14
28940	17.14	17.14	17.13	17.13	17.13	17.13	17.13	17.13	17.13	17.13
28960	17.12	17.12	17.12	17.12	17.12	17.12	17.12	17.12	17.12	17.11
28980	17.11	17.11	17.11	17.11	17.11	17.11	17.11	17.1	17.1	17.1
29000	17.1	17.1	17.1	17.1	17.1	17.1	17.09	17.09	17.09	17.09

TABLE 2-18
HALF WAVELENGTH INVERTED-VEE DIPOLE
29000 - 29500 KILOHERTZ
USE: AMATEUR

Measurements, in feet, advance in 1 kHz increments from left to right across columns.

kHz	feet	feet	feet	feet	feet	feet	feet	feet	feet	feet
29000	17.1	17.1	17.1	17.1	17.1	17.1	17.09	17.09	17.09	17.09
29020	17.09	17.09	17.09	17.09	17.08	17.08	17.08	17.08	17.08	17.08
29040	17.08	17.08	17.08	17.07	17.07	17.07	17.07	17.07	17.07	17.07
29060	17.07	17.06	17.06	17.06	17.06	17.06	17.06	17.06	17.06	17.06
29080	17.05	17.05	17.05	17.05	17.05	17.05	17.05	17.05	17.04	17.04
29100	17.04	17.04	17.04	17.04	17.04	17.04	17.04	17.03	17.03	17.03
29120	17.03	17.03	17.03	17.03	17.03	17.02	17.02	17.02	17.02	17.02
29140	17.02	17.02	17.02	17.02	17.01	17.01	17.01	17.01	17.01	17.01
29160	17.01	17.01	17.01	17	17	17	17	17	17	17
29180	17	16.99	16.99	16.99	16.99	16.99	16.99	16.99	16.99	16.99
29200	16.98	16.98	16.98	16.98	16.98	16.98	16.98	16.98	16.97	16.97
29220	16.97	16.97	16.97	16.97	16.97	16.97	16.97	16.96	16.96	16.96
29240	16.96	16.96	16.96	16.96	16.96	16.96	16.95	16.95	16.95	16.95
29260	16.95	16.95	16.95	16.95	16.94	16.94	16.94	16.94	16.94	16.94
29280	16.94	16.94	16.94	16.93	16.93	16.93	16.93	16.93	16.93	16.93
29300	16.93	16.92	16.92	16.92	16.92	16.92	16.92	16.92	16.92	16.92
29320	16.91	16.91	16.91	16.91	16.91	16.91	16.91	16.91	16.91	16.9
29340	16.9	16.9	16.9	16.9	16.9	16.9	16.9	16.89	16.89	16.89
29360	16.89	16.89	16.89	16.89	16.89	16.89	16.88	16.88	16.88	16.88
29380	16.88	16.88	16.88	16.88	16.88	16.87	16.87	16.87	16.87	16.87
29400	16.87	16.87	16.87	16.87	16.86	16.86	16.86	16.86	16.86	16.86
29420	16.86	16.86	16.85	16.85	16.85	16.85	16.85	16.85	16.85	16.85
29440	16.85	16.84	16.84	16.84	16.84	16.84	16.84	16.84	16.84	16.84
29460	16.83	16.83	16.83	16.83	16.83	16.83	16.83	16.83	16.82	16.82
29480	16.82	16.82	16.82	16.82	16.82	16.82	16.82	16.81	16.81	16.81
29500	16.81	16.81	16.81	16.81	16.81	16.81	16.8	16.8	16.8	16.8

TABLE 2-19
HALF WAVELENGTH INVERTED-VEE DIPOLE
29500 - 30000 KILOHERTZ
USE: AMATEUR

Measurements, in feet, advance in 1 kHz increments from left to right across columns.

kHz	feet	feet	feet	feet	feet	feet	feet	feet	feet	feet
29500	16.81	16.81	16.81	16.81	16.81	16.81	16.8	16.8	16.8	16.8
29520	16.8	16.8	16.8	16.8	16.8	16.79	16.79	16.79	16.79	16.79
29540	16.79	16.79	16.79	16.79	16.78	16.78	16.78	16.78	16.78	16.78
29560	16.78	16.78	16.77	16.77	16.77	16.77	16.77	16.77	16.77	16.77
29580	16.77	16.76	16.76	16.76	16.76	16.76	16.76	16.76	16.76	16.76
29600	16.75	16.75	16.75	16.75	16.75	16.75	16.75	16.75	16.75	16.74
29620	16.74	16.74	16.74	16.74	16.74	16.74	16.74	16.74	16.73	16.73
29640	16.73	16.73	16.73	16.73	16.73	16.73	16.73	16.72	16.72	16.72
29660	16.72	16.72	16.72	16.72	16.72	16.71	16.71	16.71	16.71	16.71
29680	16.71	16.71	16.71	16.71	16.7	16.7	16.7	16.7	16.7	16.7
29700	16.7	16.7	16.7	16.69	16.69	16.69	16.69	16.69	16.69	16.69
29720	16.69	16.69	16.68	16.68	16.68	16.68	16.68	16.68	16.68	16.68
29740	16.68	16.67	16.67	16.67	16.67	16.67	16.67	16.67	16.67	16.67
29760	16.66	16.66	16.66	16.66	16.66	16.66	16.66	16.66	16.66	16.65
29780	16.65	16.65	16.65	16.65	16.65	16.65	16.65	16.65	16.64	16.64
29800	16.64	16.64	16.64	16.64	16.64	16.64	16.64	16.63	16.63	16.63
29820	16.63	16.63	16.63	16.63	16.63	16.63	16.62	16.62	16.62	16.62
29840	16.62	16.62	16.62	16.62	16.62	16.61	16.61	16.61	16.61	16.61
29860	16.61	16.61	16.61	16.61	16.6	16.6	16.6	16.6	16.6	16.6
29880	16.6	16.6	16.6	16.59	16.59	16.59	16.59	16.59	16.59	16.59
29900	16.59	16.59	16.58	16.58	16.58	16.58	16.58	16.58	16.58	16.58
29920	16.58	16.57	16.57	16.57	16.57	16.57	16.57	16.57	16.57	16.57
29940	16.56	16.56	16.56	16.56	16.56	16.56	16.56	16.56	16.56	16.55
29960	16.55	16.55	16.55	16.55	16.55	16.55	16.55	16.55	16.54	16.54
29980	16.54	16.54	16.54	16.54	16.54	16.54	16.54	16.53	16.53	16.53
30000	16.53	16.53	16.53	16.53	16.53	16.53	16.52	16.52	16.52	16.52

Chapter 3
$\frac{3}{4}$-Wavelength Inverted-Vee

The term "Vee" causes confusion when used to describe an antenna. It seems that an inverted-vee dipole is not quite the same antenna as an inverted-vee beam. These antennas have different lengths and perform quite differently. The half-wavelength "inverted-vee" of Chapter 2 is little more than an attempt to reduce the space requirements needed for a regular dipole, or to obtain a lower angle of radiation. The lower angle of radiation, incidentally, is useful for working DX stations, especially in the low-frequency band. A vee beam, on the other hand, is fed from one end, rather than the center as is the inverted-vee dipole. This antenna exhibits a strong directivity, as opposed to the inverted-vee dipole, which has a directivity less than that of the regular dipole.

The charts at the end of this chapter give the lengths required to make a ¾-wavelength inverted-vee dipole, or one leg of a vee-beam (of which, more in a moment). The simple inverted-vee dipole of the last chapter used two quarter-wavelength legs, fed at the center. But the truth is that the legs may be any *odd multiple* of a quarter wavelength. Some evidence exists of better performance when the legs are longer than quarter wavelength. But note this advantage can only be obtained by those for whom the inverted-vee dipole was *not* used because of space limitations. The charts at the end of the Chapter show the lengths for each leg of the antenna. The other charts show overall length, each leg being one-half of the length shown. The legs of a ¾-wavelength inverted-vee dipole at the low end of the 80-meter band are each over 212 feet long! This means that the total antenna length will be over 400 feet, which is not exactly "small city lot" size. An advantage of this, of course, is that a low-band—80-, 75-, or 40-meter—antenna will often work on at least one other high-frequency band.

Figure 3-1 shows the diagram for a vee-beam antenna. This is a special case of the inverted-vee design, and is fed off-center. The short section is a quarter wavelength. The dimensions for this section can be found either by dividing the dimensions in the following charts by three, or by consulting one of the quarter-wavelength tables in the chapter on quarter-wavelength verticals.

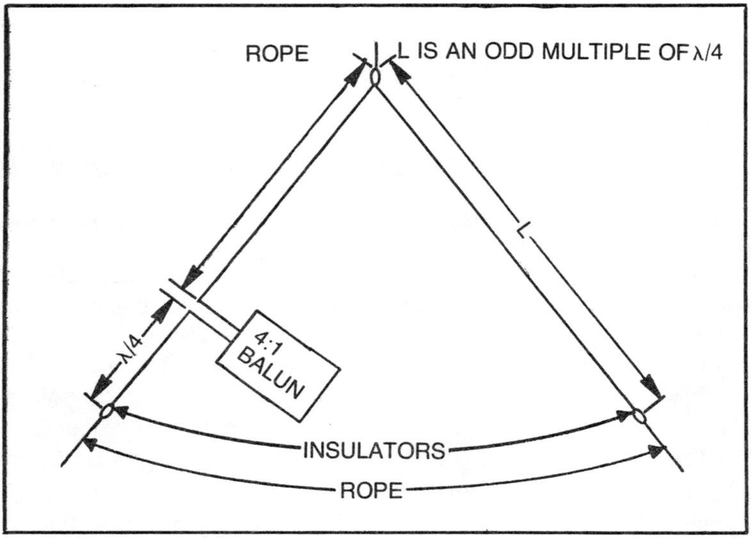

Fig. 3-1. Vee-beam (inverted).

The longer section is an odd multiple of quarter wavelength. The usual wisdom is that the "longer the better." But, once again, we must face the constraints of small lots, and choose a shorter version. The ¾-wavelength tables given in this chapter will give lengths appropriate for most cases. Note that ¾-wavelength is the next higher "odd multiple" over a simple quarter-wavelength (which would be the antenna of Chapter 2.)

Tuning is not needed on this antenna if the length is, indeed, close to an odd multiple of quarter wavelength. There is, however, a higher impedance at the feedpoint than is normally found on the half-wavelength versions of Chapter 2. This makes the use of a 4:1 balun transformer very wise. In fact, it may be found that the impedance is still mismatched with 4:1, so using one of those baluns that cover a wider range of impedance may be advisable. You should use one of the wide-band baluns wound on a toroid core if a wide frequency range of operation is anticipated. Baluns made from coaxial cable are designed for a single frequency of operation.

The center of the antenna, where the support is given, is at the center of the *entire length* of the antenna, not the center of the long section.

The directivity pattern of the antenna is shown in Fig. 3-2. Note that the maximum lobes are off the ends of the antenna. In both the half-wavelength antennas covered previously (i.e. dipole and inverted-vee dipole), the maximum radiation was perpendicular to

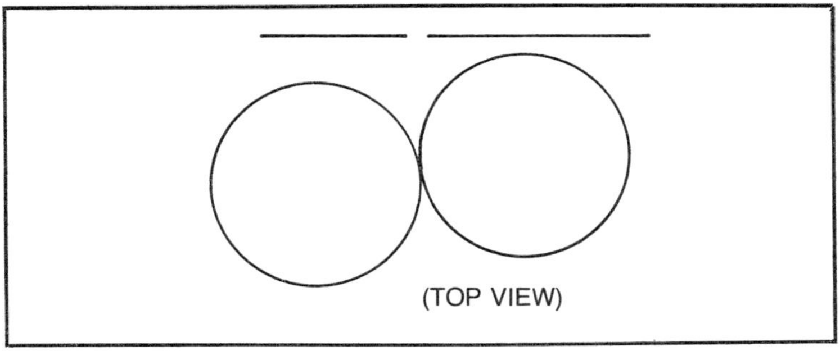

Fig. 3-2. Polar pattern of a vee-beam.

the wire. But exactly the opposite situation is found on the inverted-vee antenna.

Another type of vee beam antenna is the horizontal vee beam of Fig. 3-3. The vee-beam in this case can be viewed as a pair of long wire antennas, at an angle to each other, fed at the apex. An antenna tuner is often used at the feedpoint. For current feed, the most common type used by amateurs, the legs of the vee beam must be an odd multiple of quarter wavelength. Again, the ¾-wavelength tables may be used.

The included angle, θ, between the legs has a tremendous effect on the performance of the vee beam antenna. Figure 3-4

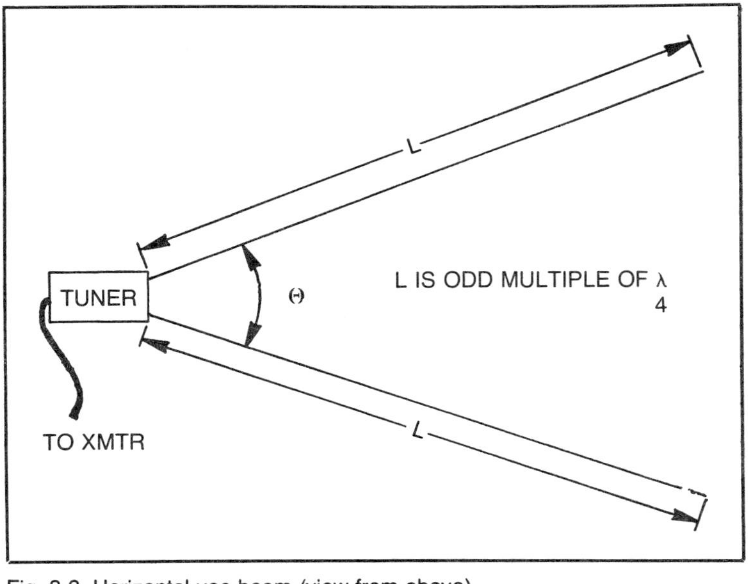

Fig. 3-3. Horizontal vee-beam (view from above).

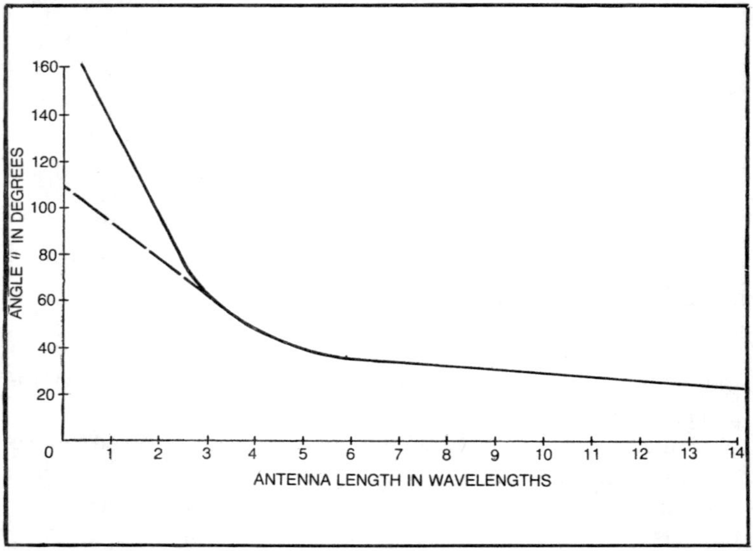

Fig. 3-4. Length-angle plot.

shows a graph of the optimum angle for radiators of various sizes. At the short lengths available to most of us, something less than two or three wavelengths, the included angle must be very high, so a compromise must be reached. This is represented by the dashed line in Fig. 3-4.

The gain of a vee-beam varies with the included angle. The gain of the vee-beam at ¾-wavelengths is little better than that of a dipole. But as the length goes up, then so does the gain. Most engineering charts on the vee-beam tell us that gains of 12 B are possible at lengths over 10 wavelengths. This accounts for the stories you may sometimes hear of the person who connects the wrong piece of coaxial cable to a two-meter rig, such that an 80-meter long wire vee-beam, or trap dipole, is used as the antenna. Tremendous signal strengths have been reported, because on 2 meters an 80-meter vee-beam has legs that approach 100 wavelengths!

TABLE 3-1
3/4-WAVELENGTH INVERTED-VEE
300 - 500 KILOHERTZ
USE: MARINE/RADIO NAVIGATION

Measurements, in feet, advance in 1 kHz increments from left to right across columns.

kHz	feet	feet	feet	feet	feet	feet	feet	feet	feet	feet
300	2480	2471.76	2463.57	2455.44	2447.36	2439.34	2431.37	2423.45	2415.58	2407.76
310	2400	2392.28	2384.61	2376.99	2369.42	2361.9	2354.43	2347	2339.62	2332.28
320	2325	2317.75	2310.55	2303.4	2296.29	2289.23	2282.2	2275.22	2268.29	2261.39
330	2254.54	2247.73	2240.96	2234.23	2227.54	2220.89	2214.28	2207.71	2201.18	2194.69
340	2188.23	2181.81	2175.43	2169.09	2162.79	2156.52	2150.28	2144.09	2137.93	2131.8
350	2125.71	2119.65	2113.63	2107.64	2101.69	2095.77	2089.88	2084.03	2078.21	2072.42
360	2066.66	2060.94	2055.24	2049.58	2043.95	2038.35	2032.78	2027.24	2021.73	2016.26
370	2010.81	2005.39	2000	1994.63	1989.3	1984	1978.72	1973.47	1968.25	1963.06
380	1957.89	1952.75	1947.64	1942.55	1937.5	1932.46	1927.46	1922.48	1917.52	1912.59
390	1907.69	1902.81	1897.95	1893.12	1888.32	1883.54	1878.78	1874.05	1869.34	1864.66
400	1860	1855.36	1850.74	1846.15	1841.58	1837.03	1832.51	1828	1823.52	1819.07
410	1814.63	1810.21	1805.82	1801.45	1797.1	1792.77	1788.46	1784.17	1779.9	1775.65
420	1771.42	1767.22	1763.03	1758.86	1754.71	1750.58	1746.47	1742.38	1738.31	1734.26
430	1730.23	1726.21	1722.22	1718.24	1714.28	1710.34	1706.42	1702.51	1698.63	1694.76
440	1690.9	1687.07	1683.25	1679.45	1675.67	1671.91	1668.16	1664.42	1660.71	1657.01
450	1653.33	1649.66	1646.01	1642.38	1638.76	1635.16	1631.57	1628	1624.45	1620.91
460	1617.39	1613.88	1610.38	1606.91	1603.44	1600	1596.56	1593.14	1589.74	1586.35
470	1582.97	1579.61	1576.27	1572.93	1569.62	1566.31	1563.02	1559.74	1556.48	1553.23
480	1550	1546.77	1543.56	1540.37	1537.19	1534.02	1530.86	1527.72	1524.59	1521.47
490	1518.36	1515.27	1512.19	1509.12	1506.07	1503.03	1500	1496.98	1493.97	1490.98
500	1488	1485.02	1482.07	1479.12	1476.19	1473.26	1470.35	1467.45	1464.56	1461.68

TABLE 3-2
3/4-WAVELENGTH INVERTED VEE
540 - 1605 KILOHERTZ
USE: AM BROADCASTING

Measurements, in feet, advance in 1 kHz increments from left to right across columns.

kHz	feet	feet	feet	feet	feet	feet	feet	feet	feet	feet
540	1377.77	1365.13	1352.72	1340.54	1328.57	1316.81	1305.26	1293.91	1282.75	1271.79
590	1261.01	1250.42	1240	1229.75	1219.67	1209.75	1200	1190.4	1180.95	1171.65
640	1162.5	1153.48	1144.61	1135.87	1127.27	1118.79	1110.44	1102.22	1094.11	1086.13
690	1078.26	1070.5	1062.85	1055.31	1047.88	1040.55	1033.33	1026.2	1019.17	1012.24
740	1005.4	998.65	992	985.43	978.94	972.54	966.23	960	953.84	947.77
790	941.77	935.84	930	924.22	918.51	912.88	907.31	901.81	896.38	891.01
840	885.71	880.47	875.29	870.17	865.11	860.11	855.17	850.28	845.45	840.67
890	835.95	831.28	826.66	822.09	817.58	813.11	808.69	804.32	800	795.72
940	791.48	787.3	783.15	779.05	775	770.98	767.01	763.07	759.18	755.32
990	751.51	747.73	744	740.29	736.63	733	729.41	725.85	722.33	718.84
1040	715.38	711.96	708.57	705.21	701.88	698.59	695.32	692.09	688.88	685.71
1090	682.56	679.45	676.36	673.3	670.27	667.26	664.28	661.33	658.4	655.5
1140	652.63	649.78	646.95	644.15	641.37	638.62	635.89	633.19	630.5	627.84
1190	625.21	622.59	620	617.42	614.87	612.34	609.83	607.34	604.87	602.42
1240	600	597.59	595.2	592.82	590.47	588.14	585.82	583.52	581.25	578.98
1290	576.74	574.51	572.3	570.11	567.93	565.77	563.63	561.5	559.39	557.3
1340	555.22	553.15	551.11	549.07	547.05	545.05	543.06	541.09	539.13	537.18
1390	535.25	533.33	531.42	529.53	527.65	525.79	523.94	522.1	520.27	518.46
1440	516.66	514.87	513.1	511.34	509.58	507.84	506.12	504.4	502.7	501.01
1490	499.32	497.65	496	494.35	492.71	491.08	489.47	487.86	486.27	484.69
1540	483.11	481.55	480	478.45	476.92	475.39	473.88	472.38	470.88	469.4
1590	467.92	466.45	465	463.55	462.11	460.68	459.25	457.84	456.44	455.04

TABLE 3-3
3/4-WAVELENGTH INVERTED VEE
1800 - 2000 KILOHERTZ
USE: AMATEUR

Measurements, in feet, advance in 1 kHz increments from left to right across columns.

kHz	feet	feet	feet	feet	feet	feet	feet	feet	feet	feet
1800	413.33	413.1	412.87	412.64	412.41	412.18	411.96	411.73	411.5	411.27
1810	411.04	410.82	410.59	410.36	410.14	409.91	409.69	409.46	409.24	409.01
1820	408.79	408.56	408.34	408.11	407.89	407.67	407.44	407.22	407	406.77
1830	406.55	406.33	406.11	405.89	405.67	405.44	405.22	405	404.78	404.56
1840	404.34	404.12	403.9	403.68	403.47	403.25	403.03	402.81	402.59	402.37
1850	402.16	401.94	401.72	401.51	401.29	401.07	400.86	400.64	400.43	400.21
1860	400	399.78	399.57	399.35	399.14	398.92	398.71	398.5	398.28	398.07
1870	397.86	397.64	397.43	397.22	397.01	396.8	396.58	396.37	396.16	395.95
1880	395.74	395.53	395.32	395.11	394.9	394.69	394.48	394.27	394.06	393.85
1890	393.65	393.44	393.23	393.02	392.81	392.61	392.4	392.19	391.99	391.78
1900	391.57	391.37	391.16	390.96	390.75	390.55	390.34	390.14	389.93	389.73
1910	389.52	389.32	389.12	388.91	388.71	388.51	388.3	388.1	387.9	387.7
1920	387.5	387.29	387.09	386.89	386.69	386.49	386.29	386.09	385.89	385.69
1930	385.49	385.29	385.09	384.89	384.69	384.49	384.29	384.09	383.9	383.7
1940	383.5	383.3	383.11	382.91	382.71	382.51	382.32	382.12	381.93	381.73
1950	381.53	381.34	381.14	380.95	380.75	380.56	380.36	380.17	379.97	379.78
1960	379.59	379.39	379.2	379.01	378.81	378.62	378.43	378.24	378.04	377.85
1970	377.66	377.47	377.28	377.09	376.89	376.7	376.51	376.32	376.13	375.94
1980	375.75	375.56	375.37	375.18	375	374.81	374.62	374.43	374.24	374.05
1990	373.86	373.68	373.49	373.3	373.11	372.93	372.74	372.55	372.37	372.18
2000	372	371.81	371.62	371.44	371.25	371.07	370.88	370.7	370.51	370.33

TABLE 3-4
3/4-WAVELENGTH INVERTED-VEE DIPOLE
3500 - 4000 KILOHERTZ
USE: AMATEUR

Measurements, in feet, advance in 1 kHz increments from left to right across columns.

kHz	feet	feet	feet	feet	feet	feet	feet	feet	feet	feet
3500	212.57	212.45	212.32	212.2	212.08	211.96	211.84	211.72	211.6	211.48
3520	211.36	211.24	211.12	211	210.88	210.76	210.64	210.52	210.4	210.28
3540	210.16	210.05	209.93	209.81	209.69	209.57	209.45	209.34	209.22	209.1
3560	208.98	208.87	208.75	208.63	208.52	208.4	208.28	208.17	208.05	207.93
3580	207.82	207.7	207.58	207.47	207.35	207.24	207.12	207.01	206.89	206.78
3600	206.66	206.55	206.43	206.32	206.2	206.09	205.98	205.86	205.75	205.63
3620	205.52	205.41	205.29	205.18	205.07	204.95	204.84	204.73	204.62	204.5
3640	204.39	204.28	204.17	204.05	203.94	203.83	203.72	203.61	203.5	203.38
3660	203.27	203.16	203.05	202.94	202.83	202.72	202.61	202.5	202.39	202.28
3680	202.17	202.06	201.95	201.84	201.73	201.62	201.51	201.4	201.29	201.18
3700	201.08	200.97	200.86	200.75	200.64	200.53	200.43	200.32	200.21	200.1
3720	200	199.89	199.78	199.67	199.57	199.46	199.35	199.25	199.14	199.03
3740	198.93	198.82	198.71	198.61	198.5	198.4	198.29	198.18	198.08	197.97
3760	197.87	197.76	197.66	197.55	197.45	197.34	197.24	197.13	197.03	196.92
3780	196.82	196.72	196.61	196.51	196.4	196.3	196.2	196.09	195.99	195.89
3800	195.78	195.68	195.58	195.48	195.37	195.27	195.17	195.07	194.96	194.86
3820	194.76	194.66	194.56	194.45	194.35	194.25	194.15	194.05	193.95	193.85
3840	193.75	193.64	193.54	193.44	193.34	193.24	193.14	193.04	192.94	192.84
3860	192.74	192.64	192.54	192.44	192.34	192.24	192.14	192.04	191.95	191.85
3880	191.75	191.65	191.55	191.45	191.35	191.25	191.16	191.06	190.96	190.86
3900	190.76	190.67	190.57	190.47	190.37	190.28	190.18	190.08	189.98	189.89
3920	189.79	189.69	189.6	189.5	189.4	189.31	189.21	189.12	189.02	188.92
3940	188.83	188.73	188.64	188.54	188.44	188.35	188.25	188.16	188.06	187.97
3960	187.87	187.78	187.68	187.59	187.5	187.4	187.31	187.21	187.12	187.02
3980	186.93	186.84	186.74	186.65	186.55	186.46	186.37	186.27	186.18	186.09
4000	186	185.9	185.81	185.72	185.62	185.53	185.44	185.35	185.25	185.16

TABLE 3-5
3/4-WAVELENGTH INVERTED-VEE DIPOLE
5500 - 6000 KILOHERTZ
USE: INT L. BROADCAST

Measurements, in feet, advance in 1 kHz increments from left to right across columns.

kHz	feet	feet	feet	feet	feet	feet	feet	feet	feet	feet
5500	135.27	135.22	135.17	135.12	135.07	135.02	134.97	134.92	134.88	134.83
5520	134.78	134.73	134.68	134.63	134.58	134.53	134.49	134.44	134.39	134.34
5540	134.29	134.24	134.19	134.15	134.1	134.05	134	133.95	133.9	133.86
5560	133.81	133.76	133.71	133.66	133.62	133.57	133.52	133.47	133.42	133.38
5580	133.33	133.28	133.23	133.19	133.14	133.09	133.04	132.99	132.95	132.9
5600	132.85	132.8	132.76	132.71	132.66	132.62	132.57	132.52	132.47	132.43
5620	132.38	132.33	132.29	132.24	132.19	132.14	132.1	132.05	132	131.96
5640	131.91	131.86	131.82	131.77	131.72	131.68	131.63	131.58	131.54	131.49
5660	131.44	131.4	131.35	131.3	131.26	131.21	131.17	131.12	131.07	131.03
5680	130.98	130.93	130.89	130.84	130.8	130.75	130.7	130.66	130.61	130.57
5700	130.52	130.48	130.43	130.38	130.34	130.29	130.25	130.2	130.16	130.11
5720	130.06	130.02	129.97	129.93	129.88	129.84	129.79	129.75	129.7	129.66
5740	129.61	129.57	129.52	129.48	129.43	129.39	129.34	129.3	129.25	129.21
5760	129.16	129.12	129.07	129.03	128.98	128.94	128.89	128.85	128.8	128.76
5780	128.71	128.67	128.63	128.58	128.54	128.49	128.45	128.4	128.36	128.32
5800	128.27	128.23	128.18	128.14	128.09	128.05	128.01	127.96	127.92	127.87
5820	127.83	127.79	127.74	127.7	127.65	127.61	127.57	127.52	127.48	127.44
5840	127.39	127.35	127.31	127.26	127.22	127.17	127.13	127.09	127.04	127
5860	126.96	126.91	126.87	126.83	126.78	126.74	126.7	126.65	126.61	126.57
5880	126.53	126.48	126.44	126.4	126.35	126.31	126.27	126.23	126.18	126.14
5900	126.1	126.05	126.01	125.97	125.93	125.88	125.84	125.8	125.76	125.71
5920	125.67	125.63	125.59	125.54	125.5	125.46	125.42	125.37	125.33	125.29
5940	125.25	125.21	125.16	125.12	125.08	125.04	125	124.95	124.91	124.87
5960	124.83	124.79	124.74	124.7	124.66	124.62	124.58	124.53	124.49	124.45
5980	124.41	124.37	124.33	124.29	124.24	124.2	124.16	124.12	124.08	124.04
6000	124	123.95	123.91	123.87	123.83	123.79	123.75	123.71	123.67	123.62

TABLE 3-6
3/4-WAVELENGTH INVERTED-VEE DIPOLE
6000 - 6500 KILOHERTZ
USE: INT'L. BROADCAST

Measurements, in feet, advance in 1 kHz increments from left to right across columns.

kHz	feet	feet	feet	feet	feet	feet	feet	feet	feet	feet
6000	124	123.95	123.91	123.87	123.83	123.79	123.75	123.71	123.67	123.62
6020	123.58	123.54	123.5	123.46	123.42	123.38	123.34	123.3	123.26	123.21
6040	123.17	123.13	123.09	123.05	123.01	122.97	122.93	122.89	122.85	122.81
6060	122.77	122.73	122.69	122.65	122.61	122.57	122.52	122.48	122.44	122.4
6080	122.36	122.32	122.28	122.24	122.2	122.16	122.12	122.08	122.04	122
6100	121.96	121.92	121.88	121.84	121.8	121.76	121.72	121.68	121.64	121.6
6120	121.56	121.52	121.48	121.44	121.4	121.37	121.33	121.29	121.25	121.21
6140	121.17	121.13	121.09	121.05	121.01	120.97	120.93	120.89	120.85	120.81
6160	120.77	120.74	120.7	120.66	120.62	120.58	120.54	120.5	120.46	120.42
6180	120.38	120.34	120.31	120.27	120.23	120.19	120.15	120.11	120.07	120.03
6200	120	119.96	119.92	119.88	119.84	119.8	119.76	119.72	119.69	119.65
6220	119.61	119.57	119.53	119.49	119.46	119.42	119.38	119.34	119.3	119.26
6240	119.23	119.19	119.15	119.11	119.07	119.04	119	118.96	118.92	118.88
6260	118.84	118.81	118.77	118.73	118.69	118.66	118.62	118.58	118.54	118.5
6280	118.47	118.43	118.39	118.35	118.32	118.28	118.24	118.2	118.17	118.13
6300	118.09	118.05	118.02	117.98	117.94	117.9	117.87	117.83	117.79	117.75
6320	117.72	117.68	117.64	117.6	117.57	117.53	117.49	117.46	117.42	117.38
6340	117.35	117.31	117.27	117.23	117.2	117.16	117.12	117.09	117.05	117.01
6360	116.98	116.94	116.9	116.87	116.83	116.79	116.76	116.72	116.68	116.65
6380	116.61	116.57	116.54	116.5	116.46	116.43	116.39	116.35	116.32	116.28
6400	116.25	116.21	116.17	116.14	116.1	116.06	116.03	115.99	115.96	115.92
6420	115.88	115.85	115.81	115.77	115.74	115.7	115.67	115.63	115.59	115.56
6440	115.52	115.49	115.45	115.42	115.38	115.34	115.31	115.27	115.24	115.2
6460	115.17	115.13	115.09	115.06	115.02	114.99	114.95	114.92	114.88	114.85
6480	114.81	114.77	114.74	114.7	114.67	114.63	114.6	114.56	114.53	114.49
6500	114.46	114.42	114.39	114.35	114.32	114.28	114.25	114.21	114.18	114.14

TABLE 3-7
3/4-WAVELENGTH INVERTED-VEE DIPOLE
7000 - 7300 KILOHERTZ
USE: AMATEUR

Measurements, in feet, advance in 1 kHz increments from left to right across columns.

kHz	feet	feet	feet	feet	feet	feet	feet	feet	feet	feet
7000	106.28	106.25	106.22	106.19	106.16	106.13	106.1	106.07	106.04	106.01
7020	105.98	105.95	105.92	105.89	105.86	105.83	105.8	105.77	105.74	105.71
7040	105.68	105.65	105.62	105.59	105.56	105.53	105.5	105.47	105.44	105.41
7060	105.38	105.35	105.32	105.29	105.26	105.23	105.2	105.17	105.14	105.11
7080	105.08	105.05	105.02	104.99	104.96	104.93	104.9	104.87	104.84	104.81
7100	104.78	104.75	104.72	104.7	104.67	104.64	104.61	104.58	104.55	104.52
7120	104.49	104.46	104.43	104.4	104.37	104.34	104.31	104.28	104.26	104.23
7140	104.2	104.17	104.14	104.11	104.08	104.05	104.02	103.99	103.96	103.93
7160	103.91	103.88	103.85	103.82	103.79	103.76	103.73	103.7	103.67	103.65
7180	103.62	103.59	103.56	103.53	103.5	103.47	103.44	103.41	103.39	103.36
7200	103.33	103.3	103.27	103.24	103.21	103.19	103.16	103.13	103.1	103.07
7220	103.04	103.01	102.99	102.96	102.93	102.9	102.87	102.84	102.81	102.79
7240	102.76	102.73	102.7	102.67	102.64	102.62	102.59	102.56	102.53	102.5
7260	102.47	102.45	102.42	102.39	102.36	102.33	102.31	102.28	102.25	102.22
7280	102.19	102.16	102.14	102.11	102.08	102.05	102.02	102	101.97	101.94
7300	101.91	101.88	101.86	101.83	101.8	101.77	101.75	101.72	101.69	101.66

TABLE 3-8
3/4-WAVELENGTH INVERTED-VEE DIPOLE
9500 - 10000 KILOHERTZ
USE: INT'L. BROADCAST

Measurements, in feet, advance in 1 kHz increments from left to right across columns.

kHz	feet	feet	feet	feet	feet	feet	feet	feet	feet	feet
9500	78.31	78.29	78.26	78.26	78.24	78.23	78.21	78.2	78.18	78.16
9520	78.15	78.13	78.11	78.1	78.08	78.06	78.05	78.03	78.02	78
9540	77.98	77.97	77.95	77.93	77.92	77.9	77.88	77.87	77.85	77.84
9560	77.82	77.8	77.79	77.77	77.75	77.74	77.72	77.71	77.69	77.67
9580	77.66	77.64	77.62	77.61	77.59	77.58	77.56	77.54	77.53	77.51
9600	77.5	77.48	77.46	77.45	77.43	77.41	77.4	77.38	77.37	77.35
9620	77.33	77.32	77.3	77.29	77.27	77.25	77.24	77.22	77.21	77.19
9640	77.17	77.16	77.14	77.13	77.11	77.09	77.08	77.06	77.05	77.03
9660	77.01	77	76.98	76.97	76.95	76.93	76.92	76.9	76.89	76.87
9680	76.85	76.84	76.82	76.81	76.79	76.78	76.76	76.74	76.73	76.71
9700	76.7	76.68	76.66	76.65	76.63	76.62	76.6	76.59	76.57	76.55
9720	76.54	76.52	76.51	76.49	76.48	76.46	76.44	76.43	76.41	76.4
9740	76.38	76.37	76.35	76.33	76.32	76.3	76.29	76.27	76.26	76.24
9760	76.22	76.21	76.19	76.18	76.16	76.15	76.13	76.12	76.1	76.08
9780	76.07	76.05	76.04	76.02	76.01	75.99	75.98	75.96	75.94	75.93
9800	75.91	75.9	75.88	75.87	75.85	75.84	75.82	75.81	75.79	75.77
9820	75.76	75.74	75.73	75.71	75.7	75.68	75.67	75.65	75.64	75.62
9840	75.6	75.59	75.57	75.56	75.54	75.53	75.51	75.5	75.48	75.47
9860	75.45	75.44	75.42	75.41	75.39	75.37	75.36	75.34	75.33	75.31
9880	75.3	75.28	75.27	75.25	75.24	75.22	75.21	75.19	75.18	75.16
9900	75.15	75.13	75.12	75.1	75.09	75.07	75.06	75.04	75.03	75.01
9920	75	74.98	74.96	74.95	74.93	74.92	74.9	74.89	74.87	74.86
9940	74.84	74.83	74.81	74.8	74.78	74.77	74.75	74.74	74.72	74.71
9960	74.69	74.68	74.66	74.65	74.63	74.62	74.6	74.59	74.57	74.56
9980	74.54	74.53	74.51	74.5	74.48	74.47	74.45	74.44	74.42	74.41
10000	74.4	74.38	74.37	74.35	74.34	74.32	74.31	74.29	74.28	74.26

TABLE 3-9
3/4-WAVELENGTH INVERTED-VEE DIPOLE
11500 - 12000 KILOHERTZ
USE: INT'L. BROADCAST

Measurements, in feet, advance in 1 kHz increments from left to right across columns.

kHz	feet	feet	feet	feet	feet	feet	feet	feet	feet	feet
11500	64.69	64.68	64.67	64.66	64.65	64.63	64.62	64.61	64.6	64.59
11520	64.58	64.57	64.56	64.54	64.53	64.52	64.51	64.5	64.49	64.48
11540	64.47	64.46	64.44	64.43	64.42	64.41	64.4	64.39	64.38	64.37
11560	64.35	64.34	64.33	64.32	64.31	64.3	64.29	64.28	64.27	64.25
11580	64.24	64.23	64.22	64.21	64.2	64.19	64.18	64.17	64.16	64.14
11600	64.13	64.12	64.11	64.1	64.09	64.08	64.07	64.06	64.04	64.03
11620	64.02	64.01	64	63.99	63.98	63.97	63.96	63.95	63.93	63.92
11640	63.91	63.9	63.89	63.88	63.87	63.86	63.85	63.84	63.82	63.81
11660	63.8	63.79	63.78	63.77	63.76	63.75	63.74	63.73	63.72	63.7
11680	63.69	63.68	63.67	63.66	63.65	63.64	63.63	63.62	63.61	63.6
11700	63.58	63.57	63.56	63.55	63.54	63.53	63.52	63.51	63.5	63.49
11720	63.48	63.47	63.45	63.44	63.43	63.42	63.41	63.4	63.39	63.38
11740	63.37	63.36	63.35	63.34	63.32	63.31	63.3	63.29	63.28	63.27
11760	63.26	63.25	63.24	63.23	63.22	63.21	63.2	63.19	63.17	63.16
11780	63.15	63.14	63.13	63.12	63.11	63.1	63.09	63.08	63.07	63.06
11800	63.05	63.04	63.02	63.01	63	62.99	62.98	62.97	62.96	62.95
11820	62.94	62.93	62.92	62.91	62.9	62.89	62.88	62.86	62.85	62.84
11840	62.83	62.82	62.81	62.8	62.79	62.78	62.77	62.76	62.75	62.74
11860	62.73	62.72	62.71	62.7	62.68	62.67	62.66	62.65	62.64	62.63
11880	62.62	62.61	62.6	62.59	62.58	62.57	62.56	62.55	62.54	62.53
11900	62.52	62.51	62.5	62.48	62.47	62.46	62.45	62.44	62.43	62.42
11920	62.41	62.4	62.39	62.38	62.37	62.36	62.35	62.34	62.33	62.32
11940	62.31	62.3	62.29	62.28	62.26	62.25	62.24	62.23	62.22	62.21
11960	62.2	62.19	62.18	62.17	62.16	62.15	62.14	62.13	62.12	62.11
11980	62.1	62.09	62.08	62.07	62.06	62.05	62.04	62.03	62.02	62.01
12000	62	61.98	61.97	61.96	61.95	61.94	61.93	61.92	61.91	61.9

TABLE 3-10
3/4-WAVELENGTH INVERTED-VEE DIPOLE
14000 - 14450 KILOHERTZ
USE: AMATEUR

Measurements, in feet, advance in 1 kHz increments from left to right across columns.

kHz	feet	feet	feet	feet	feet	feet	feet	feet	feet	feet
14000	53.14	53.13	53.12	53.12	53.11	53.1	53.09	53.08	53.08	53.07
14020	53.06	53.05	53.05	53.04	53.03	53.02	53.02	53.01	53	52.99
14040	52.99	52.98	52.97	52.96	52.96	52.95	52.94	52.93	52.93	52.92
14060	52.91	52.9	52.9	52.89	52.88	52.87	52.87	52.86	52.85	52.84
14080	52.84	52.83	52.82	52.81	52.81	52.8	52.79	52.78	52.78	52.77
14100	52.76	52.75	52.75	52.74	52.73	52.72	52.72	52.71	52.7	52.69
14120	52.69	52.68	52.67	52.66	52.66	52.65	52.64	52.63	52.63	52.62
14140	52.61	52.6	52.6	52.59	52.58	52.57	52.57	52.56	52.55	52.54
14160	52.54	52.53	52.52	52.52	52.51	52.5	52.49	52.49	52.48	52.47
14180	52.46	52.46	52.45	52.44	52.43	52.43	52.42	52.41	52.4	52.4
14200	52.39	52.38	52.37	52.37	52.36	52.35	52.35	52.34	52.33	52.32
14220	52.32	52.31	52.3	52.29	52.29	52.28	52.27	52.26	52.26	52.25
14240	52.24	52.23	52.23	52.22	52.21	52.21	52.2	52.19	52.18	52.18
14260	52.17	52.16	52.15	52.15	52.14	52.13	52.13	52.12	52.11	52.1
14280	52.1	52.09	52.08	52.07	52.07	52.06	52.05	52.04	52.04	52.03
14300	52.02	52.02	52.01	52	51.99	51.99	51.98	51.97	51.96	51.96
14320	51.95	51.94	51.94	51.93	51.92	51.91	51.91	51.9	51.89	51.89
14340	51.88	51.87	51.86	51.86	51.85	51.84	51.83	51.83	51.82	51.81
14360	51.81	51.8	51.79	51.78	51.78	51.77	51.76	51.76	51.75	51.74
14380	51.73	51.73	51.72	51.71	51.7	51.7	51.69	51.68	51.68	51.67
14400	51.66	51.65	51.65	51.64	51.63	51.63	51.62	51.61	51.6	51.6
14420	51.59	51.58	51.58	51.57	51.56	51.55	51.55	51.54	51.53	51.53
14440	51.52	51.51	51.5	51.5	51.49	51.48	51.48	51.47	51.46	51.45

TABLE 3-11
3/4-WAVELENGTH INVERTED-VEE DIPOLE
15500 - 16000 KILOHERTZ
USE: INT'L. BROADCAST

Measurements, in feet, advance in 1 kHz increments from left to right across columns.

kHz	feet	feet	feet	feet	feet	feet	feet	feet	feet	feet
15500	48	47.99	47.98	47.98	47.97	47.96	47.96	47.95	47.95	47.94
15520	47.93	47.93	47.92	47.91	47.91	47.9	47.9	47.89	47.88	47.88
15540	47.87	47.87	47.86	47.85	47.85	47.84	47.83	47.83	47.82	47.82
15560	47.81	47.8	47.8	47.79	47.79	47.78	47.77	47.77	47.76	47.75
15580	47.75	47.74	47.74	47.73	47.72	47.72	47.71	47.71	47.7	47.69
15600	47.69	47.68	47.68	47.67	47.66	47.66	47.65	47.64	47.64	47.63
15620	47.63	47.62	47.61	47.61	47.6	47.6	47.59	47.58	47.58	47.57
15640	47.57	47.56	47.55	47.55	47.54	47.53	47.53	47.52	47.52	47.51
15660	47.5	47.5	47.49	47.49	47.48	47.47	47.47	47.46	47.46	47.45
15680	47.44	47.44	47.43	47.43	47.42	47.41	47.41	47.4	47.4	47.39
15700	47.38	47.38	47.37	47.37	47.36	47.35	47.35	47.34	47.34	47.33
15720	47.32	47.32	47.31	47.31	47.3	47.29	47.29	47.28	47.28	47.27
15740	47.26	47.26	47.25	47.25	47.24	47.23	47.23	47.22	47.22	47.21
15760	47.2	47.2	47.19	47.19	47.18	47.17	47.17	47.16	47.16	47.15
15780	47.14	47.14	47.13	47.13	47.12	47.11	47.11	47.1	47.1	47.09
15800	47.08	47.08	47.07	47.07	47.06	47.05	47.05	47.04	47.04	47.03
15820	47.02	47.02	47.01	47.01	47	46.99	46.99	46.98	46.98	46.97
15840	46.96	46.96	46.95	46.95	46.94	46.94	46.93	46.92	46.92	46.91
15860	46.91	46.9	46.89	46.89	46.88	46.88	46.87	46.86	46.86	46.85
15880	46.85	46.84	46.83	46.83	46.82	46.82	46.81	46.81	46.8	46.79
15900	46.79	46.78	46.78	46.77	46.76	46.76	46.75	46.75	46.74	46.73
15920	46.73	46.72	46.72	46.71	46.71	46.7	46.69	46.69	46.68	46.68
15940	46.67	46.66	46.66	46.65	46.65	46.64	46.63	46.63	46.62	46.62
15960	46.61	46.61	46.6	46.59	46.59	46.58	46.58	46.57	46.56	46.56
15980	46.55	46.55	46.54	46.54	46.53	46.52	46.52	46.51	46.51	46.5
16000	46.5	46.49	46.48	46.48	46.47	46.47	46.46	46.45	46.45	46.44

TABLE 3-12
3/4-WAVELENGTH INVERTED-VEE DIPOLE
16500 - 17000 KILOHERTZ
USE: INT´L. BROADCAST

Measurements, in feet, advance in 1 kHz increments from left to right across columns.

kHz	feet	feet	feet	feet	feet	feet	feet	feet	feet	feet
16500	45.09	45.08	45.07	45.07	45.06	45.06	45.05	45.05	45.04	45.04
16520	45.03	45.03	45.02	45.01	45.01	45	45	44.99	44.99	44.98
16540	44.98	44.97	44.97	44.96	44.96	44.95	44.94	44.94	44.93	44.93
16560	44.92	44.92	44.91	44.91	44.9	44.9	44.89	44.88	44.88	44.87
16580	44.87	44.86	44.86	44.85	44.85	44.84	44.84	44.83	44.83	44.82
16600	44.81	44.81	44.8	44.8	44.79	44.79	44.78	44.78	44.77	44.77
16620	44.76	44.75	44.75	44.74	44.74	44.73	44.73	44.72	44.72	44.71
16640	44.71	44.7	44.7	44.69	44.69	44.68	44.67	44.67	44.66	44.66
16660	44.65	44.65	44.64	44.64	44.63	44.63	44.62	44.62	44.61	44.6
16680	44.6	44.59	44.59	44.58	44.58	44.57	44.57	44.56	44.56	44.55
16700	44.55	44.54	44.54	44.53	44.52	44.52	44.51	44.51	44.5	44.5
16720	44.49	44.49	44.48	44.48	44.47	44.47	44.46	44.46	44.45	44.44
16740	44.44	44.43	44.43	44.42	44.42	44.41	44.41	44.4	44.4	44.39
16760	44.39	44.38	44.38	44.37	44.37	44.36	44.35	44.35	44.34	44.34
16780	44.33	44.33	44.32	44.32	44.31	44.31	44.3	44.3	44.29	44.29
16800	44.28	44.28	44.27	44.26	44.26	44.25	44.25	44.24	44.24	44.23
16820	44.23	44.22	44.22	44.21	44.21	44.2	44.2	44.19	44.19	44.18
16840	44.18	44.17	44.17	44.16	44.15	44.15	44.14	44.14	44.13	44.13
16860	44.12	44.12	44.11	44.11	44.1	44.1	44.09	44.09	44.08	44.08
16880	44.07	44.07	44.06	44.06	44.05	44.04	44.04	44.03	44.03	44.02
16900	44.02	44.01	44.01	44	44	43.99	43.99	43.98	43.98	43.97
16920	43.97	43.96	43.96	43.95	43.95	43.94	43.94	43.93	43.93	43.92
16940	43.91	43.91	43.9	43.9	43.89	43.89	43.88	43.88	43.87	43.87
16960	43.86	43.86	43.85	43.85	43.84	43.84	43.83	43.83	43.82	43.82
16980	43.81	43.81	43.8	43.8	43.79	43.79	43.78	43.78	43.77	43.76
17000	43.76	43.75	43.75	43.74	43.74	43.73	43.73	43.72	43.72	43.71

TABLE 3-13
3/4-WAVELENGTH INVERTED-VEE DIPOLE
18500 - 19000 KILOHERTZ
USE: INT'L. BROADCAST

Measurements, in feet, advance in 1 kHz increments from left to right across columns.

kHz	feet	feet	feet	feet	feet	feet	feet	feet	feet	feet
18500	40.21	40.21	40.2	40.2	40.19	40.19	40.19	40.18	40.18	40.17
18520	40.17	40.16	40.16	40.15	40.15	40.15	40.14	40.14	40.13	40.13
18540	40.12	40.12	40.12	40.11	40.11	40.1	40.1	40.09	40.09	40.09
18560	40.08	40.08	40.07	40.07	40.06	40.06	40.06	40.05	40.05	40.04
18580	40.04	40.03	40.03	40.03	40.02	40.02	40.01	40.01	40	40
18600	40	39.99	39.99	39.98	39.98	39.97	39.97	39.96	39.96	39.96
18620	39.95	39.95	39.94	39.94	39.93	39.93	39.93	39.92	39.92	39.91
18640	39.91	39.9	39.9	39.9	39.89	39.89	39.88	39.88	39.87	39.87
18660	39.87	39.86	39.86	39.85	39.85	39.85	39.84	39.84	39.83	39.83
18680	39.82	39.82	39.82	39.81	39.81	39.8	39.8	39.79	39.79	39.79
18700	39.78	39.78	39.77	39.77	39.76	39.76	39.76	39.75	39.75	39.74
18720	39.74	39.73	39.73	39.73	39.72	39.72	39.71	39.71	39.7	39.7
18740	39.7	39.69	39.69	39.68	39.68	39.68	39.67	39.67	39.66	39.66
18760	39.65	39.65	39.65	39.64	39.64	39.63	39.63	39.62	39.62	39.62
18780	39.61	39.61	39.6	39.6	39.59	39.59	39.59	39.58	39.58	39.57
18800	39.57	39.57	39.56	39.56	39.55	39.55	39.54	39.54	39.54	39.53
18820	39.53	39.52	39.52	39.51	39.51	39.51	39.5	39.5	39.49	39.49
18840	39.49	39.48	39.48	39.47	39.47	39.46	39.46	39.46	39.45	39.45
18860	39.44	39.44	39.44	39.43	39.43	39.42	39.42	39.41	39.41	39.41
18880	39.4	39.4	39.39	39.39	39.39	39.38	39.38	39.37	39.37	39.36
18900	39.36	39.36	39.35	39.35	39.34	39.34	39.34	39.33	39.33	39.32
18920	39.32	39.31	39.31	39.31	39.3	39.3	39.29	39.29	39.29	39.28
18940	39.28	39.27	39.27	39.26	39.26	39.26	39.25	39.25	39.24	39.24
18960	39.24	39.23	39.23	39.22	39.22	39.21	39.21	39.21	39.2	39.2
18980	39.19	39.19	39.19	39.18	39.18	39.17	39.17	39.17	39.16	39.16
19000	39.15	39.15	39.14	39.14	39.14	39.13	39.13	39.12	39.12	39.12

TABLE 3-14
3/4-WAVELENGTH INVERTED-VEE DIPOLE
21000 - 21450 KILOHERTZ
USE: AMATEUR

Measurements, in feet, advance in 1 kHz increments from left to right across columns.

kHz	feet	feet	feet	feet	feet	feet	feet	feet	feet	feet
21000	35.42	35.42	35.42	35.41	35.41	35.41	35.4	35.4	35.4	35.39
21020	35.39	35.39	35.38	35.38	35.38	35.37	35.37	35.37	35.36	35.36
21040	35.36	35.35	35.35	35.35	35.34	35.34	35.34	35.33	35.33	35.33
21060	35.32	35.32	35.32	35.31	35.31	35.31	35.3	35.3	35.3	35.29
21080	35.29	35.29	35.28	35.28	35.28	35.27	35.27	35.27	35.26	35.26
21100	35.26	35.25	35.25	35.25	35.24	35.24	35.24	35.23	35.23	35.23
21120	35.22	35.22	35.22	35.21	35.21	35.21	35.2	35.2	35.2	35.19
21140	35.19	35.19	35.18	35.18	35.18	35.17	35.17	35.17	35.16	35.16
21160	35.16	35.15	35.15	35.15	35.14	35.14	35.14	35.13	35.13	35.13
21180	35.12	35.12	35.12	35.11	35.11	35.11	35.1	35.1	35.1	35.09
21200	35.09	35.09	35.08	35.08	35.08	35.07	35.07	35.07	35.06	35.06
21220	35.06	35.05	35.05	35.05	35.04	35.04	35.04	35.03	35.03	35.03
21240	35.02	35.02	35.02	35.01	35.01	35.01	35	35	35	34.99
21260	34.99	34.99	34.98	34.98	34.98	34.97	34.97	34.97	34.96	34.96
21280	34.96	34.95	34.95	34.95	34.94	34.94	34.94	34.93	34.93	34.93
21300	34.92	34.92	34.92	34.91	34.91	34.91	34.9	34.9	34.9	34.9
21320	34.89	34.89	34.89	34.88	34.88	34.88	34.87	34.87	34.87	34.86
21340	34.86	34.86	34.85	34.85	34.85	34.84	34.84	34.84	34.83	34.83
21360	34.83	34.82	34.82	34.82	34.81	34.81	34.81	34.8	34.8	34.8
21380	34.79	34.79	34.79	34.78	34.78	34.78	34.77	34.77	34.77	34.76
21400	34.76	34.76	34.75	34.75	34.75	34.75	34.74	34.74	34.74	34.73
21420	34.73	34.73	34.72	34.72	34.72	34.71	34.71	34.71	34.7	34.7
21440	34.7	34.69	34.69	34.69	34.68	34.68	34.68	34.67	34.67	34.67

TABLE 3-15
3/4-WAVELENGTH INVERTED-VEE DIPOLE
25500 - 26000 KILOHERTZ
USE: INT'L. BROADCAST

Measurements, in feet, advance in 1 kHz increments from left to right across columns.

kHz	feet	feet	feet	feet	feet	feet	feet	feet	feet	feet
25500	29.17	29.17	29.17	29.16	29.16	29.16	29.16	29.16	29.15	29.15
25520	29.15	29.15	29.14	29.14	29.14	29.14	29.13	29.13	29.13	29.13
25540	29.13	29.12	29.12	29.12	29.12	29.11	29.11	29.11	29.11	29.11
25560	29.1	29.1	29.1	29.1	29.09	29.09	29.09	29.09	29.08	29.08
25580	29.08	29.08	29.08	29.07	29.07	29.07	29.07	29.06	29.06	29.06
25600	29.06	29.06	29.05	29.05	29.05	29.05	29.04	29.04	29.04	29.04
25620	29.03	29.03	29.03	29.03	29.03	29.02	29.02	29.02	29.02	29.01
25640	29.01	29.01	29.01	29.01	29	29	29	29	28.99	28.99
25660	28.99	28.99	28.99	28.98	28.98	28.98	28.98	28.97	28.97	28.97
25680	28.97	28.96	28.96	28.96	28.96	28.96	28.95	28.95	28.95	28.95
25700	28.94	28.94	28.94	28.94	28.94	28.93	28.93	28.93	28.93	28.92
25720	28.92	28.92	28.92	28.92	28.91	28.91	28.91	28.91	28.9	28.9
25740	28.9	28.9	28.89	28.89	28.89	28.89	28.89	28.88	28.88	28.88
25760	28.88	28.87	28.87	28.87	28.87	28.87	28.86	28.86	28.86	28.86
25780	28.85	28.85	28.85	28.85	28.85	28.84	28.84	28.84	28.84	28.83
25800	28.83	28.83	28.83	28.83	28.82	28.82	28.82	28.82	28.81	28.81
25820	28.81	28.81	28.81	28.8	28.8	28.8	28.8	28.79	28.79	28.79
25840	28.79	28.79	28.78	28.78	28.78	28.78	28.77	28.77	28.77	28.77
25860	28.77	28.76	28.76	28.76	28.76	28.75	28.75	28.75	28.75	28.75
25880	28.74	28.74	28.74	28.74	28.73	28.73	28.73	28.73	28.73	28.72
25900	28.72	28.72	28.72	28.71	28.71	28.71	28.71	28.71	28.7	28.7
25920	28.7	28.7	28.69	28.69	28.69	28.69	28.69	28.68	28.68	28.68
25940	28.68	28.67	28.67	28.67	28.67	28.67	28.66	28.66	28.66	28.66
25960	28.65	28.65	28.65	28.65	28.65	28.64	28.64	28.64	28.64	28.63
25980	28.63	28.63	28.63	28.63	28.62	28.62	28.62	28.62	28.61	28.61
26000	28.61	28.61	28.61	28.6	28.6	28.6	28.6	28.59	28.59	28.59

TABLE 3-16
3/4-WAVELENGTH INVERTED-VEE DIPOLE
26500 - 27000 KILOHERTZ
USE: INT'L. BROADCAST

Measurements, in feet, advance in 1 kHz increments from left to right across columns.

kHz	feet	feet	feet	feet	feet	feet	feet	feet	feet	feet
26500	28.07	28.07	28.07	28.06	28.06	28.06	28.06	28.06	28.05	28.05
26520	28.05	28.05	28.05	28.04	28.04	28.04	28.04	28.03	28.03	28.03
26540	28.03	28.03	28.02	28.02	28.02	28.02	28.02	28.01	28.01	28.01
26560	28.01	28	28	28	28	28	27.99	27.99	27.99	27.99
26580	27.99	27.98	27.98	27.98	27.98	27.98	27.97	27.97	27.97	27.97
26600	27.96	27.96	27.96	27.96	27.96	27.95	27.95	27.95	27.95	27.95
26620	27.94	27.94	27.94	27.94	27.94	27.93	27.93	27.93	27.93	27.93
26640	27.92	27.92	27.92	27.92	27.91	27.91	27.91	27.91	27.91	27.9
26660	27.9	27.9	27.9	27.9	27.89	27.89	27.89	27.89	27.89	27.88
26680	27.88	27.88	27.88	27.87	27.87	27.87	27.87	27.87	27.86	27.86
26700	27.86	27.86	27.86	27.85	27.85	27.85	27.85	27.85	27.84	27.84
26720	27.84	27.84	27.84	27.83	27.83	27.83	27.83	27.82	27.82	27.82
26740	27.82	27.82	27.81	27.81	27.81	27.81	27.81	27.8	27.8	27.8
26760	27.8	27.8	27.79	27.79	27.79	27.79	27.79	27.78	27.78	27.78
26780	27.78	27.77	27.77	27.77	27.77	27.77	27.76	27.76	27.76	27.76
26800	27.76	27.75	27.75	27.75	27.75	27.75	27.74	27.74	27.74	27.74
26820	27.74	27.73	27.73	27.73	27.73	27.73	27.72	27.72	27.72	27.72
26840	27.71	27.71	27.71	27.71	27.71	27.7	27.7	27.7	27.7	27.7
26860	27.69	27.69	27.69	27.69	27.69	27.68	27.68	27.68	27.68	27.68
26880	27.67	27.67	27.67	27.67	27.67	27.66	27.66	27.66	27.66	27.66
26900	27.65	27.65	27.65	27.65	27.64	27.64	27.64	27.64	27.64	27.63
26920	27.63	27.63	27.63	27.63	27.62	27.62	27.62	27.62	27.62	27.61
26940	27.61	27.61	27.61	27.61	27.6	27.6	27.6	27.6	27.6	27.59
26960	27.59	27.59	27.59	27.59	27.58	27.58	27.58	27.58	27.58	27.57
26980	27.57	27.57	27.57	27.56	27.56	27.56	27.56	27.56	27.55	27.55
27000	27.55	27.55	27.55	27.54	27.54	27.54	27.54	27.54	27.53	27.53

TABLE 3-17
3/4-WAVELENGTH INVERTED-VEE DIPOLE
28500 - 29000 KILOHERTZ
USE: AMATEUR

Measurements, in feet, advance in 1 kHz increments from left to right across columns.

kHz	feet	feet	feet	feet	feet	feet	feet	feet	feet	feet
28500	26.1	26.1	26.1	26.09	26.09	26.09	26.09	26.09	26.09	26.08
28520	26.08	26.08	26.08	26.08	26.07	26.07	26.07	26.07	26.07	26.07
28540	26.06	26.06	26.06	26.06	26.06	26.05	26.05	26.05	26.05	26.05
28560	26.05	26.04	26.04	26.04	26.04	26.04	26.03	26.03	26.03	26.03
28580	26.03	26.03	26.02	26.02	26.02	26.02	26.02	26.01	26.01	26.01
28600	26.01	26.01	26.01	26	26	26	26	26	25.99	25.99
28620	25.99	25.99	25.99	25.99	25.98	25.98	25.98	25.98	25.98	25.97
28640	25.97	25.97	25.97	25.97	25.97	25.96	25.96	25.96	25.96	25.96
28660	25.95	25.95	25.95	25.95	25.95	25.95	25.94	25.94	25.94	25.94
28680	25.94	25.93	25.93	25.93	25.93	25.93	25.93	25.92	25.92	25.92
28700	25.92	25.92	25.91	25.91	25.91	25.91	25.91	25.91	25.9	25.9
28720	25.9	25.9	25.9	25.89	25.89	25.89	25.89	25.89	25.89	25.88
28740	25.88	25.88	25.88	25.88	25.88	25.87	25.87	25.87	25.87	25.87
28760	25.86	25.86	25.86	25.86	25.86	25.86	25.85	25.85	25.85	25.85
28780	25.85	25.84	25.84	25.84	25.84	25.84	25.84	25.83	25.83	25.83
28800	25.83	25.83	25.82	25.82	25.82	25.82	25.82	25.82	25.81	25.81
28820	25.81	25.81	25.81	25.81	25.8	25.8	25.8	25.8	25.8	25.79
28840	25.79	25.79	25.79	25.79	25.79	25.78	25.78	25.78	25.78	25.78
28860	25.77	25.77	25.77	25.77	25.77	25.77	25.76	25.76	25.76	25.76
28880	25.76	25.75	25.75	25.75	25.75	25.75	25.75	25.74	25.74	25.74
28900	25.74	25.74	25.74	25.73	25.73	25.73	25.73	25.73	25.72	25.72
28920	25.72	25.72	25.72	25.72	25.71	25.71	25.71	25.71	25.71	25.71
28940	25.7	25.7	25.7	25.7	25.7	25.69	25.69	25.69	25.69	25.69
28960	25.69	25.68	25.68	25.68	25.68	25.68	25.67	25.67	25.67	25.67
28980	25.67	25.67	25.66	25.66	25.66	25.66	25.66	25.66	25.65	25.65
29000	25.65	25.65	25.65	25.64	25.64	25.64	25.64	25.64	25.64	25.63

TABLE 3-18
3/4-WAVELENGTH INVERTED-VEE
29000 - 29500 KILOHERTZ
USE: AMATEUR

Measurements, in feet, advance in 1 kHz increments from left to right across columns.

kHz	feet	feet	feet	feet	feet	feet	feet	feet	feet	feet
29000	24.2	24.2	24.2	24.2	24.2	24.19	24.19	24.19	24.19	24.19
29020	24.19	24.18	24.18	24.18	24.18	24.18	24.18	24.17	24.17	24.17
29040	24.17	24.17	24.17	24.16	24.16	24.16	24.16	24.16	24.16	24.15
29060	24.15	24.15	24.15	24.15	24.15	24.14	24.14	24.14	24.14	24.14
29080	24.14	24.13	24.13	24.13	24.13	24.13	24.13	24.12	24.12	24.12
29100	24.12	24.12	24.12	24.11	24.11	24.11	24.11	24.11	24.11	24.1
29120	24.1	24.1	24.1	24.1	24.1	24.09	24.09	24.09	24.09	24.09
29140	24.09	24.08	24.08	24.08	24.08	24.08	24.08	24.07	24.07	24.07
29160	24.07	24.07	24.07	24.06	24.06	24.06	24.06	24.06	24.06	24.05
29180	24.05	24.05	24.05	24.05	24.05	24.04	24.04	24.04	24.04	24.04
29200	24.04	24.03	24.03	24.03	24.03	24.03	24.03	24.02	24.02	24.02
29220	24.02	24.02	24.02	24.01	24.01	24.01	24.01	24.01	24.01	24
29240	24	24	24	24	24	24	23.99	23.99	23.99	23.99
29260	23.99	23.99	23.98	23.98	23.98	23.98	23.98	23.98	23.97	23.97
29280	23.97	23.97	23.97	23.97	23.96	23.96	23.96	23.96	23.96	23.96
29300	23.95	23.95	23.95	23.95	23.95	23.95	23.94	23.94	23.94	23.94
29320	23.94	23.94	23.93	23.93	23.93	23.93	23.93	23.93	23.92	23.92
29340	23.92	23.92	23.92	23.92	23.91	23.91	23.91	23.91	23.91	23.91
29360	23.91	23.9	23.9	23.9	23.9	23.9	23.9	23.89	23.89	23.89
29380	23.89	23.89	23.89	23.88	23.88	23.88	23.88	23.88	23.88	23.87
29400	23.87	23.87	23.87	23.87	23.87	23.86	23.86	23.86	23.86	23.86
29420	23.86	23.85	23.85	23.85	23.85	23.85	23.85	23.84	23.84	23.84
29440	23.84	23.84	23.84	23.84	23.83	23.83	23.83	23.83	23.83	23.83
29460	23.82	23.82	23.82	23.82	23.82	23.82	23.81	23.81	23.81	23.81
29480	23.81	23.81	23.8	23.8	23.8	23.8	23.8	23.8	23.79	23.79
29500	23.79	23.79	23.79	23.79	23.79	23.78	23.78	23.78	23.78	23.78

TABLE 3-19
3/4-WAVELENGTH INVERTED-VEE DIPOLE
29500 - 30000 KILOHERTZ
USE: AMATEUR

Measurements, in feet, advance in 1 kHz increments from left to right across columns.

kHz	feet	feet	feet	feet	feet	feet	feet	feet	feet	feet
29500	25.22	25.21	25.21	25.21	25.21	25.21	25.21	25.2	25.2	25.2
29520	25.2	25.2	25.19	25.19	25.19	25.19	25.19	25.19	25.18	25.18
29540	25.18	25.18	25.18	25.18	25.17	25.17	25.17	25.17	25.17	25.17
29560	25.16	25.16	25.16	25.16	25.16	25.16	25.15	25.15	25.15	25.15
29580	25.15	25.15	25.14	25.14	25.14	25.14	25.14	25.14	25.13	25.13
29600	25.13	25.13	25.13	25.13	25.12	25.12	25.12	25.12	25.12	25.11
29620	25.11	25.11	25.11	25.11	25.11	25.1	25.1	25.1	25.1	25.1
29640	25.1	25.09	25.09	25.09	25.09	25.09	25.09	25.08	25.08	25.08
29660	25.08	25.08	25.08	25.07	25.07	25.07	25.07	25.07	25.07	25.06
29680	25.06	25.06	25.06	25.06	25.06	25.05	25.05	25.05	25.05	25.05
29700	25.05	25.04	25.04	25.04	25.04	25.04	25.04	25.03	25.03	25.03
29720	25.03	25.03	25.03	25.02	25.02	25.02	25.02	25.02	25.02	25.01
29740	25.01	25.01	25.01	25.01	25.01	25	25	25	25	25
29760	25	24.99	24.99	24.99	24.99	24.99	24.98	24.98	24.98	24.98
29780	24.98	24.98	24.97	24.97	24.97	24.97	24.97	24.97	24.96	24.96
29800	24.96	24.96	24.96	24.96	24.95	24.95	24.95	24.95	24.95	24.95
29820	24.94	24.94	24.94	24.94	24.94	24.94	24.93	24.93	24.93	24.93
29840	24.93	24.93	24.92	24.92	24.92	24.92	24.92	24.92	24.91	24.91
29860	24.91	24.91	24.91	24.91	24.9	24.9	24.9	24.9	24.9	24.9
29880	24.89	24.89	24.89	24.89	24.89	24.89	24.88	24.88	24.88	24.88
29900	24.88	24.88	24.87	24.87	24.87	24.87	24.87	24.87	24.86	24.86
29920	24.86	24.86	24.86	24.86	24.85	24.85	24.85	24.85	24.85	24.85
29940	24.84	24.84	24.84	24.84	24.84	24.84	24.83	24.83	24.83	24.83
29960	24.83	24.83	24.82	24.82	24.82	24.82	24.82	24.82	24.81	24.81
29980	24.81	24.81	24.81	24.81	24.8	24.8	24.8	24.8	24.8	24.8
30000	24.8	24.79	24.79	24.79	24.79	24.79	24.79	24.78	24.78	24.78

Chapter 4
Quarter-Wavelength Verticals

Next to the half-wavelength dipole, the vertical antenna is one of the most popular amateur high-frequency antennas. It is relatively easy to build and is easy to "tame," or get working. Figure 4-1A shows the basic format for a vertical antenna. A quarter-wavelength radiator is fed from a piece of coaxial cable. The inside conductor of the transmission line is connected to the vertical radiator element, and the outer conductor—the shield—is connected to the ground. Some texts claim the missing half wavelength is inside the ground but this is merely a poor attempt to describe the action.

Shown in Fig. 4-1B are the graphs of the current and voltage distribution patterns along the antenna. As in the half-wavelength dipole, the current at the feedpoint is maximum, and the voltage minimum. This arrangement is considered the best practice in most cases.

The nominal feedpoint impedance of the vertical antenna is 50 ohms, so ordinary 52-ohm coaxial cable, such as RG-58 and RG-8, makes a good match. In reality, however, we find that the feed-point impedance of the vertical antenna may be anything from 5 ohms to 75 ohms! Of course, this could create quite a mismatch to 52-ohm coaxial cable. Some companies, such as Palomar Engineers, offer multiple-impedance transformers made especially for matching the vertical to 52-ohm coaxial cable.

The normal arrangement for a ground-mounted vertical radiator is shown in Fig. 4-2. The coaxial cable is brought to the base of the antenna, and the center conductor is connected to the antenna element. The shield of the cable is connected to earth ground. But very few locations have an adequate earth ground. In fact, most people have very poor soil conductivity—consequently, antennas—so something must be done to improve the situation.

One method for improving the earth is to drive one or preferably more antenna ground rods (available from most amateur and commercial suppliers) into the earth right at the base of the antenna. The short TV-type ground rods are not very effective, so use the long types. If you can find, and afford, the power company and

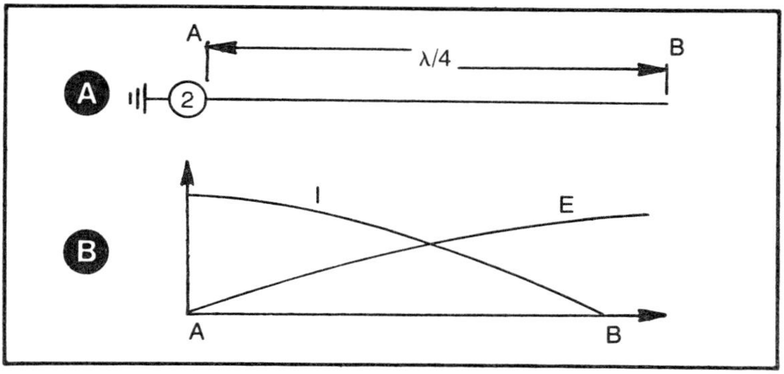

Fig. 4-1. Quarter wavelength vertical at (A) with I/E distribution at (B).

telephone type of ground rods, use them. These are 0.75 inches in diameter, and over 8 feet long. The more buried the ground rod is, the better the ground connection will be.

Some people use the ground rods, and then treat the soil with copper sulphate. This supposedly improves the electrical conductivity of the soil.

Another tactic is to connect quarter-wavelength radials—lengths of copper wire that are each a quarter-wavelength long—to the shield and the ground to form a ground plane. The radials are stretched out from the point where the radiator is located, symmetrically around the antenna site. Be warned, however, that if the radials are on the surface, they may trip (or shock) someone, and that may leave you liable to suit. A better approach seems to be burial of the radials underground. You might think this to be a terrible chore, at first (it is, really), but it is not as bad as most people believe. The radials do not have to be buried deep in a trench, only a slit a few inches deep. The tip of a shovel or garden spade will slice the soil sufficiently to bury the radial.

Fig. 4-2. Direct feed with coaxial cable (ground installation).

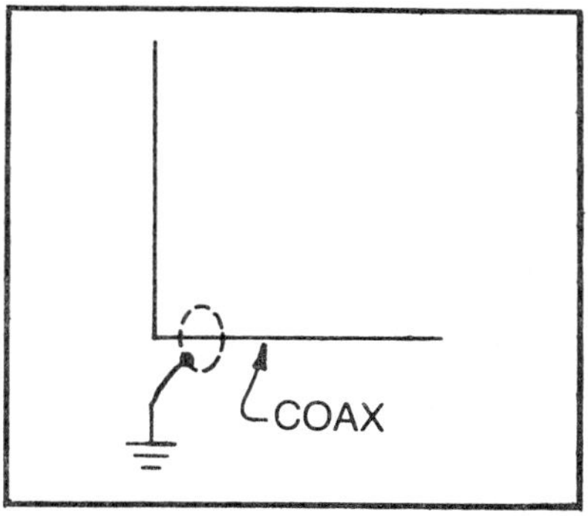

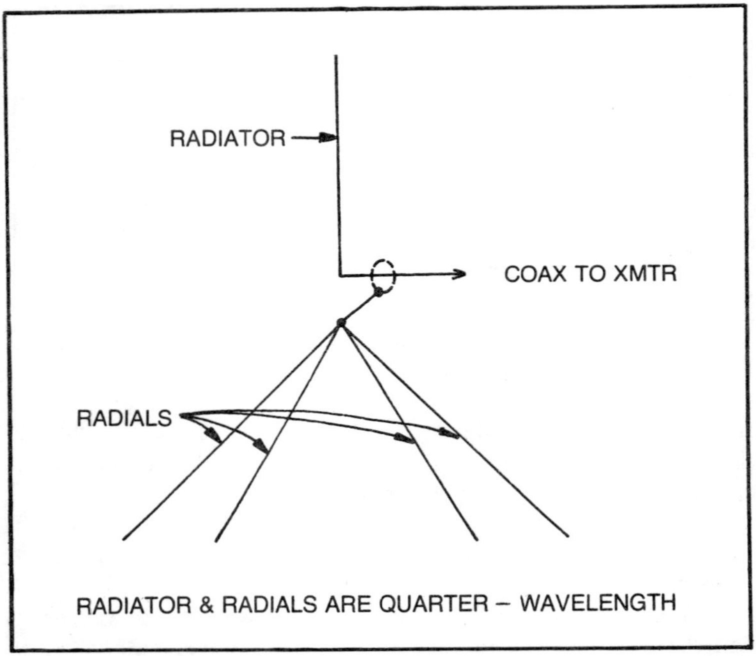

Fig. 4-3. Direct feed with coaxial cable (above ground installation).

Another approach is to visit the site of your new home (assuming, of course, that you can afford a new home at today's prices!) before the sod is put down and bury conductors. Two friends of mine solved this in different ways. One unrolled many yards of copper screen wire all over the part of his homesite where the antenna was to be located, forming an in-site ground plane. The other chap, the chief engineer at a small AM radio station was concerned about the ground conductivity of the station site. When the station was being built, and the site was still ungrassed, he went out with what looked like four zillion miles of copper wire and laid down a grid pattern, as well as radials, on the soil. The wires were interconnected at the cross points and at the antenna site. This formed a very effective ground plane. But before you rush out and buy wire or screening, consider the cost. The AM radio station spent over $5000 a decade ago for that wire. Fortunately, most amateurs can be satisfied with the results of only a few radials buried a few inches deep using the tip of a shovel.

In many cases, it is not possible or desirable to mount a vertical antenna on the ground. Figure 4-3 shows how radials may also be used to form a ground plane for a vertical antenna.

The radials are positioned symmetrically about the antenna (if possible). If the radials form a 90° angle with the radiator, the impedance will be low and some form of matching will be needed. Also, the angle of radiation will be higher. But if the radials are

Fig. 4-4. Quickie vertical for 20-meters.

allowed to droop at some angle greater than 90° to the radiator, the impedance will be closer to 50 ohms, and the angle of radiation is lower.

The rule for radials is make them one-quarter wavelength, and the more the merrier: six radials are better than one. The radials are usually constructed of wire, although aluminum tubing may also be used if you are a glutton for punishment.

The easiest way to make the vertical radiator element is by using aluminum tubing. Note well that *adjacent* sizes of aluminum tubing have inside and outside diameters designed so that the smaller hardware store aluminum displays have non-adjacent sizes (i. e. 0.75 and 1.0 in.), and use lengths that are too short for most amateur antenna applications. Some amateur suppliers, however, stock 10-foot lengths of adjacent size aluminum tubing. If your local supplier does not, ask why. Then, once your complaint is made, go to a metal supplier and try to get them to sell you a couple of pieces. Most of these companies are wholesalers and wouldn't know what to do when some ham comes in and orders one piece each of two sizes.

One can also obtain aluminum stock from CB antennas. Most vertical CB antennas use ⅝-wavelength radiators, which for 27 MHz is about 22 feet. If you cut the CB radials off and replace them with wire radials cut for an amateur band, then their vertical radiator may be cut to size for the amateur band. Consult the charts at the back of this chapter for the proper lengths. A 22-foot piece of aluminum can be used on any band above 14 MHz, if properly cut to length.

Figure 4-4 shows a sloppy vertical built by me, at practically no cost (less than $5!). The aluminum mast used for the vertical radiator is clamped to a piece of 2 x 4 lumber that was salvaged from another job. The piece of lumber was bolted to the cinder block wall of my undersized duplex house with molylag bolts. A tuning coil (almost invisible) at the base could be wired into the circuit by leaning out the bathroom window. This would make the 20-meter antenna into a 40-meter antenna. The radials come off at approximately a 120° angle from the radiator.

This antenna was not the best that I have ever owned, but it had one endearing property: it allowed me to get back on the air for very little money. This antenna was installed right after we bought the house, and all of our money had been given to the settlement attorney. In short, we were broke! This antenna would go wild on rainy days, and in the "blizzard of '79" (three years after it was built) it came down in a pile of snow. But in the meantime, I contacted stations in all continents and many states using my old Heath HW-101 transceiver.

TABLE 4-1
QUARTER WAVELENGTH VERTICAL
300 - 500 KILOHERTZ
USE: MARINE/RADIO NAVIGATION

Measurements, in feet, advance in 1 kHz increments from left to right across columns.

kHz	feet	feet	feet	feet	feet	feet	feet	feet	feet	feet
300	780	777.4	774.83	772.27	769.73	767.21	764.7	762.21	759.74	757.28
310	754.83	752.41	750	747.6	745.22	742.85	740.5	738.17	735.84	733.54
320	731.25	728.97	726.7	724.45	722.22	720	717.79	715.59	713.41	711.24
330	709.09	706.94	704.81	702.7	700.59	698.5	696.42	694.36	692.3	690.26
340	688.23	686.21	684.21	682.21	680.23	678.26	676.3	674.35	672.41	670.48
350	668.57	666.66	664.77	662.88	661.01	659.15	657.3	655.46	653.63	651.81
360	650	648.19	646.4	644.62	642.85	641.09	639.34	637.6	635.86	634.14
370	632.43	630.72	629.03	627.34	625.66	624	622.34	620.68	619.04	617.41
380	615.78	614.17	612.56	610.96	609.37	607.79	606.21	604.65	603.09	601.54
390	600	598.46	596.93	595.41	593.9	592.4	590.9	589.42	587.93	586.46
400	585	583.54	582.08	580.64	579.2	577.77	576.35	574.93	573.52	572.12
410	570.73	569.34	567.96	566.58	565.21	563.85	562.5	561.15	559.8	558.47
420	557.14	555.81	554.5	553.19	551.88	550.58	549.29	548	546.72	545.45
430	544.18	542.92	541.66	540.41	539.17	537.93	536.69	535.46	534.24	533.02
440	531.81	530.61	529.41	528.21	527.02	525.84	524.66	523.48	522.32	521.15
450	520	518.84	517.69	516.55	515.41	514.28	513.15	512.03	510.91	509.8
460	508.69	507.59	506.49	505.39	504.31	503.22	502.14	501.07	500	498.93
470	497.87	496.81	495.76	494.71	493.67	492.63	491.59	490.56	489.53	488.51
480	487.5	486.48	485.47	484.47	483.47	482.47	481.48	480.49	479.5	478.52
490	477.55	476.57	475.6	474.64	473.68	472.72	471.77	470.82	469.87	468.93
500	468	467.06	466.13	465.2	464.28	463.36	462.45	461.53	460.62	459.72

TABLE 4-2
QUARTER WAVELENGTH VERTICAL
540 - 1605 KILOHERTZ
USE: AM BROADCASTING

Measurements, in feet, advance in 1 kHz increments from left to right across columns.

kHz	feet	feet	feet	feet	feet	feet	feet	feet	feet	feet
540	433.33	429.35	425.45	421.62	417.85	414.15	410.52	406.95	403.44	400
590	396.61	393.27	390	386.77	383.6	380.48	377.41	374.4	371.42	368.5
640	365.62	362.79	360	357.25	354.54	351.87	349.25	346.66	344.11	341.6
690	339.13	336.69	334.28	331.91	329.57	327.27	325	322.75	320.54	318.36
740	316.21	314.09	312	309.93	307.89	305.88	303.89	301.93	300	298.08
790	296.2	294.33	292.5	290.68	288.88	287.11	285.36	283.63	281.92	280.23
840	278.57	276.92	275.29	273.68	272.09	270.52	268.96	267.42	265.9	264.4
890	262.92	261.45	260	258.56	257.14	255.73	254.34	252.97	251.61	250.26
940	248.93	247.61	246.31	245.02	243.75	242.48	241.23	240	238.77	237.56
990	236.36	235.17	234	232.83	231.68	230.54	229.41	228.29	227.18	226.08
1040	225	223.92	222.85	221.8	220.75	219.71	218.69	217.67	216.66	215.66
1090	214.67	213.69	212.72	211.76	210.81	209.86	208.92	208	207.07	206.16
1140	205.26	204.36	203.47	202.59	201.72	200.85	200	199.14	198.3	197.46
1190	196.63	195.81	195	194.19	193.38	192.59	191.8	191.02	190.24	189.47
1240	188.7	187.95	187.2	186.45	185.71	184.98	184.25	183.52	182.81	182.1
1290	181.39	180.69	180	179.31	178.62	177.94	177.27	176.6	175.93	175.28
1340	174.62	173.97	173.33	172.69	172.05	171.42	170.8	170.18	169.56	168.95
1390	168.34	167.74	167.14	166.54	165.95	165.37	164.78	164.21	163.63	163.06
1440	162.5	161.93	161.37	160.82	160.27	159.72	159.18	158.64	158.1	157.57
1490	157.04	156.52	156	155.48	154.96	154.45	153.94	153.44	152.94	152.44
1540	151.94	151.45	150.96	150.48	150	149.52	149.04	148.57	148.1	147.63
1590	147.16	146.7	146.25	145.79	145.34	144.89	144.44	144	143.55	143.11

TABLE 4-3
QUARTER WAVELENGTH VERTICAL
1800 - 2000 KILOHERTZ
USE: AMATEUR

Measurements, in feet, advance in 1 kHz increments from left to right across columns.

kHz	feet	feet	feet	feet	feet	feet	feet	feet	feet	feet
1800	130	129.92	129.85	129.78	129.71	129.63	129.56	129.49	129.42	129.35
1810	129.28	129.21	129.13	129.06	128.99	128.92	128.85	128.78	128.71	128.64
1820	128.57	128.5	128.43	128.35	128.28	128.21	128.14	128.07	128	127.93
1830	127.86	127.79	127.72	127.65	127.58	127.52	127.45	127.38	127.31	127.24
1840	127.17	127.1	127.03	126.96	126.89	126.82	126.76	126.69	126.62	126.55
1850	126.48	126.41	126.34	126.28	126.21	126.14	126.07	126	125.94	125.87
1860	125.8	125.73	125.67	125.6	125.53	125.46	125.4	125.33	125.26	125.2
1870	125.13	125.06	125	124.93	124.86	124.8	124.73	124.66	124.6	124.53
1880	124.46	124.4	124.33	124.26	124.2	124.13	124.07	124	123.94	123.87
1890	123.8	123.74	123.67	123.61	123.54	123.48	123.41	123.35	123.28	123.22
1900	123.15	123.09	123.02	122.96	122.89	122.83	122.77	122.7	122.64	122.57
1910	122.51	122.44	122.38	122.32	122.25	122.19	122.12	122.06	122	121.93
1920	121.87	121.81	121.74	121.68	121.62	121.55	121.49	121.43	121.36	121.3
1930	121.24	121.18	121.11	121.05	120.99	120.93	120.86	120.8	120.74	120.68
1940	120.61	120.55	120.49	120.43	120.37	120.3	120.24	120.18	120.12	120.06
1950	120	119.93	119.87	119.81	119.75	119.69	119.63	119.57	119.5	119.44
1960	119.38	119.32	119.26	119.2	119.14	119.08	119.02	118.96	118.9	118.84
1970	118.78	118.72	118.66	118.6	118.54	118.48	118.42	118.36	118.3	118.24
1980	118.18	118.12	118.06	118	117.94	117.88	117.82	117.76	117.7	117.64
1990	117.58	117.52	117.46	117.41	117.35	117.29	117.23	117.17	117.11	117.05
2000	117	116.94	116.88	116.82	116.76	116.7	116.65	116.59	116.53	116.47

TABLE 4-4
QUARTER WAVELENGTH VERTICAL
3500 - 4000 KILOHERTZ
USE: AMATEUR

Measurements, in feet, advance in 1 kHz increments from left to right across columns.

kHz	feet	feet	feet	feet	feet	feet	feet	feet	feet	feet
3500	66.85	66.81	66.78	66.74	66.7	66.66	66.62	66.59	66.55	66.51
3520	66.47	66.43	66.4	66.36	66.32	66.28	66.25	66.21	66.17	66.13
3540	66.1	66.06	66.02	65.98	65.95	65.91	65.87	65.84	65.8	65.76
3560	65.73	65.69	65.65	65.61	65.58	65.54	65.5	65.47	65.43	65.39
3580	65.36	65.32	65.29	65.25	65.21	65.18	65.14	65.1	65.07	65.03
3600	65	64.96	64.92	64.89	64.85	64.81	64.78	64.74	64.71	64.67
3620	64.64	64.6	64.56	64.53	64.49	64.46	64.42	64.39	64.35	64.32
3640	64.28	64.25	64.21	64.17	64.14	64.1	64.07	64.03	64	63.96
3660	63.93	63.89	63.86	63.82	63.79	63.76	63.72	63.69	63.65	63.62
3680	63.58	63.55	63.51	63.48	63.44	63.41	63.38	63.34	63.31	63.27
3700	63.24	63.2	63.17	63.14	63.1	63.07	63.03	63	62.97	62.93
3720	62.9	62.86	62.83	62.8	62.76	62.73	62.7	62.66	62.63	62.6
3740	62.56	62.53	62.5	62.46	62.43	62.4	62.36	62.33	62.3	62.26
3760	62.23	62.2	62.16	62.13	62.1	62.06	62.03	62	61.97	61.93
3780	61.9	61.87	61.83	61.8	61.77	61.74	61.7	61.67	61.64	61.61
3800	61.57	61.54	61.51	61.48	61.44	61.41	61.38	61.35	61.32	61.28
3820	61.25	61.22	61.19	61.16	61.12	61.09	61.06	61.03	61	60.96
3840	60.93	60.9	60.87	60.84	60.81	60.77	60.74	60.71	60.68	60.65
3860	60.62	60.59	60.55	60.52	60.49	60.46	60.43	60.4	60.37	60.34
3880	60.3	60.27	60.24	60.21	60.18	60.15	60.12	60.09	60.06	60.03
3900	60	59.96	59.93	59.9	59.87	59.84	59.81	59.78	59.75	59.72
3920	59.69	59.66	59.63	59.6	59.57	59.54	59.51	59.48	59.45	59.42
3940	59.39	59.36	59.33	59.3	59.27	59.24	59.21	59.18	59.15	59.12
3960	59.09	59.06	59.03	59	58.97	58.94	58.91	58.88	58.85	58.82
3980	58.79	58.76	58.73	58.7	58.67	58.64	58.61	58.58	58.55	58.52
4000	58.5	58.47	58.44	58.41	58.38	58.35	58.32	58.29	58.26	58.23

TABLE 4-5
QUARTER WAVELENGTH VERTICAL
5500 - 6000 KILOHERTZ
USE: INT'L. BROADCAST

Measurements, in feet, advance in 1 kHz increments from left to right across columns.

kHz	feet	feet	feet	feet	feet	feet	feet	feet	feet	feet
5500	42.54	42.52	42.51	42.49	42.48	42.46	42.45	42.43	42.42	42.4
5520	42.39	42.37	42.36	42.34	42.32	42.31	42.29	42.28	42.26	42.25
5540	42.23	42.22	42.2	42.19	42.17	42.16	42.14	42.13	42.11	42.1
5560	42.08	42.07	42.05	42.04	42.02	42.01	41.99	41.98	41.96	41.95
5580	41.93	41.92	41.9	41.89	41.87	41.86	41.84	41.83	41.81	41.8
5600	41.78	41.77	41.75	41.74	41.72	41.71	41.69	41.68	41.66	41.65
5620	41.63	41.62	41.6	41.59	41.57	41.56	41.54	41.53	41.51	41.5
5640	41.48	41.47	41.45	41.44	41.43	41.41	41.4	41.38	41.37	41.35
5660	41.34	41.32	41.31	41.29	41.28	41.26	41.25	41.24	41.22	41.21
5680	41.19	41.18	41.16	41.15	41.13	41.12	41.11	41.09	41.08	41.06
5700	41.05	41.03	41.02	41	40.99	40.98	40.96	40.95	40.93	40.92
5720	40.9	40.89	40.88	40.86	40.85	40.83	40.82	40.8	40.79	40.78
5740	40.76	40.75	40.73	40.72	40.7	40.69	40.68	40.66	40.65	40.63
5760	40.62	40.61	40.59	40.58	40.56	40.55	40.54	40.52	40.51	40.49
5780	40.48	40.47	40.45	40.44	40.42	40.41	40.4	40.38	40.37	40.35
5800	40.34	40.33	40.31	40.3	40.28	40.27	40.26	40.24	40.23	40.22
5820	40.2	40.19	40.17	40.16	40.15	40.13	40.12	40.1	40.09	40.08
5840	40.06	40.05	40.04	40.02	40.01	40	39.98	39.97	39.95	39.94
5860	39.93	39.91	39.9	39.89	39.87	39.86	39.85	39.83	39.82	39.8
5880	39.79	39.78	39.76	39.75	39.74	39.72	39.71	39.7	39.68	39.67
5900	39.66	39.64	39.63	39.62	39.6	39.59	39.58	39.56	39.55	39.54
5920	39.52	39.51	39.5	39.48	39.47	39.46	39.44	39.43	39.42	39.4
5940	39.39	39.38	39.36	39.35	39.34	39.32	39.31	39.3	39.28	39.27
5960	39.26	39.24	39.23	39.22	39.2	39.19	39.18	39.16	39.15	39.14
5980	39.13	39.11	39.1	39.09	39.07	39.06	39.05	39.03	39.02	39.01
6000	39	38.98	38.97	38.96	30.94	38.93	38.92	38.9	38.89	38.88

TABLE 4-6
QUARTER WAVELENGTH VERTICAL
6000 - 6500 KILOHERTZ
USE: INT'L. BROADCAST

Measurements, in feet, advance in 1 kHz increments from left to right across columns.

kHz	feet	feet	feet	feet	feet	feet	feet	feet	feet	feet
6000	39	38.98	38.97	38.96	38.94	38.93	38.92	38.9	38.89	38.88
6020	38.87	38.85	38.84	38.83	38.81	38.8	38.79	38.78	38.76	38.75
6040	38.74	38.72	38.71	38.7	38.69	38.67	38.66	38.65	38.63	38.62
6060	38.61	38.6	38.58	38.57	38.56	38.55	38.53	38.52	38.51	38.49
6080	38.48	38.47	38.46	38.44	38.43	38.42	38.41	38.39	38.38	38.37
6100	38.36	38.34	38.33	38.32	38.31	38.29	38.28	38.27	38.26	38.24
6120	38.23	38.22	38.21	38.19	38.18	38.17	38.16	38.14	38.13	38.12
6140	38.11	38.09	38.08	38.07	38.06	38.04	38.03	38.02	38.01	37.99
6160	37.98	37.97	37.96	37.95	37.93	37.92	37.91	37.9	37.88	37.87
6180	37.86	37.85	37.83	37.82	37.81	37.8	37.79	37.77	37.76	37.75
6200	37.74	37.72	37.71	37.7	37.69	37.68	37.66	37.65	37.64	37.63
6220	37.62	37.6	37.59	37.58	37.57	37.56	37.54	37.53	37.52	37.51
6240	37.5	37.48	37.47	37.46	37.45	37.44	37.42	37.41	37.4	37.39
6260	37.38	37.36	37.35	37.34	37.33	37.32	37.3	37.29	37.28	37.27
6280	37.26	37.24	37.23	37.22	37.21	37.2	37.19	37.17	37.16	37.15
6300	37.14	37.13	37.11	37.1	37.09	37.08	37.07	37.06	37.04	37.03
6320	37.02	37.01	37	36.99	36.97	36.96	36.95	36.94	36.93	36.92
6340	36.9	36.89	36.88	36.87	36.86	36.85	36.83	36.82	36.81	36.8
6360	36.79	36.78	36.76	36.75	36.74	36.73	36.72	36.71	36.7	36.68
6380	36.67	36.66	36.65	36.64	36.63	36.61	36.6	36.59	36.58	36.57
6400	36.56	36.55	36.53	36.52	36.51	36.5	36.49	36.48	36.47	36.45
6420	36.44	36.43	36.42	36.41	36.4	36.39	36.38	36.36	36.35	36.34
6440	36.33	36.32	36.31	36.3	36.29	36.27	36.26	36.25	36.24	36.23
6460	36.22	36.21	36.2	36.18	36.17	36.16	36.15	36.14	36.13	36.12
6480	36.11	36.09	36.08	36.07	36.06	36.05	36.04	36.03	36.02	36.01
6500	36	35.98	35.97	35.96	35.95	35.94	35.93	35.92	35.91	35.9

TABLE 4-7
QUARTER WAVELENGTH VERTICAL
7000 - 7300 KILOHERTZ
USE: AMATEUR

Measurements, in feet, advance in 1 kHz increments from left to right across columns.

kHz	feet	feet	feet	feet	feet	feet	feet	feet	feet	feet
7000	33.42	33.41	33.4	33.39	33.39	33.38	33.37	33.36	33.35	33.34
7020	33.33	33.32	33.31	33.3	33.29	33.28	33.27	33.26	33.25	33.24
7040	33.23	33.22	33.21	33.21	33.2	33.19	33.18	33.17	33.16	33.15
7060	33.14	33.13	33.12	33.11	33.1	33.09	33.08	33.07	33.06	33.06
7080	33.05	33.04	33.03	33.02	33.01	33	32.99	32.98	32.97	32.96
7100	32.95	32.94	32.93	32.92	32.92	32.91	32.9	32.89	32.88	32.87
7120	32.86	32.85	32.84	32.83	32.82	32.81	32.8	32.8	32.79	32.78
7140	32.77	32.76	32.75	32.74	32.73	32.72	32.71	32.7	32.69	32.69
7160	32.68	32.67	32.66	32.65	32.64	32.63	32.62	32.61	32.6	32.59
7180	32.59	32.58	32.57	32.56	32.55	32.54	32.53	32.52	32.51	32.5
7200	32.5	32.49	32.48	32.47	32.46	32.45	32.44	32.43	32.42	32.41
7220	32.4	32.4	32.39	32.38	32.37	32.36	32.35	32.34	32.33	32.32
7240	32.32	32.31	32.3	32.29	32.28	32.27	32.26	32.25	32.24	32.24
7260	32.23	32.22	32.21	32.2	32.19	32.18	32.17	32.16	32.16	32.15
7280	32.14	32.13	32.12	32.11	32.1	32.09	32.08	32.08	32.07	32.06
7300	32.05	32.04	32.03	32.02	32.01	32.01	32	31.99	31.98	31.97

TABLE 4-8
QUARTER WAVELENGTH VERTICAL
9500 - 10000 KILOHERTZ
USE: INT'L. BROADCAST

Measurements, in feet, advance in 1 kHz increments from left to right across columns.

kHz	feet	feet	feet	feet	feet	feet	feet	feet	feet	feet
9500	24.63	24.62	24.62	24.61	24.61	24.6	24.6	24.59	24.59	24.58
9520	24.57	24.57	24.56	24.56	24.55	24.55	24.54	24.54	24.53	24.53
9540	24.52	24.52	24.51	24.51	24.5	24.5	24.49	24.49	24.48	24.48
9560	24.47	24.47	24.46	24.46	24.45	24.45	24.44	24.44	24.43	24.43
9580	24.42	24.42	24.41	24.41	24.4	24.4	24.39	24.39	24.38	24.38
9600	24.37	24.36	24.36	24.35	24.35	24.34	24.34	24.33	24.33	24.32
9620	24.32	24.31	24.31	24.3	24.3	24.29	24.29	24.28	24.28	24.27
9640	24.27	24.26	24.26	24.25	24.25	24.24	24.24	24.23	24.23	24.22
9660	24.22	24.21	24.21	24.2	24.2	24.19	24.19	24.18	24.18	24.17
9680	24.17	24.16	24.16	24.15	24.15	24.14	24.14	24.13	24.13	24.12
9700	24.12	24.11	24.11	24.1	24.1	24.09	24.09	24.08	24.08	24.07
9720	24.07	24.06	24.06	24.05	24.05	24.04	24.04	24.03	24.03	24.02
9740	24.02	24.01	24.01	24	24	24	23.99	23.99	23.98	23.98
9760	23.97	23.97	23.96	23.96	23.95	23.95	23.94	23.94	23.93	23.93
9780	23.92	23.92	23.91	23.91	23.9	23.9	23.89	23.89	23.88	23.88
9800	23.87	23.87	23.86	23.86	23.85	23.85	23.84	23.84	23.83	23.83
9820	23.82	23.82	23.81	23.81	23.8	23.8	23.79	23.79	23.79	23.78
9840	23.78	23.77	23.77	23.76	23.76	23.75	23.75	23.74	23.74	23.73
9860	23.73	23.72	23.72	23.71	23.71	23.7	23.7	23.69	23.69	23.68
9880	23.68	23.67	23.67	23.66	23.66	23.66	23.65	23.65	23.64	23.64
9900	23.63	23.63	23.62	23.62	23.61	23.61	23.6	23.6	23.59	23.59
9920	23.58	23.58	23.57	23.57	23.56	23.56	23.56	23.55	23.55	23.54
9940	23.54	23.53	23.53	23.52	23.52	23.51	23.51	23.5	23.5	23.49
9960	23.49	23.48	23.48	23.47	23.47	23.47	23.46	23.46	23.45	23.45
9980	23.44	23.44	23.43	23.43	23.42	23.42	23.41	23.41	23.4	23.4
10000	23.4	23.39	23.39	23.38	23.38	23.37	23.37	23.36	23.36	23.35

TABLE 4-9
QUARTER WAVELENGTH VERTICAL
11500 - 12000 KILOHERTZ
USE: INT'L. BROADCAST

Measurements, in feet, advance in 1 kHz increments from left to right across columns.

kHz	feet	feet	feet	feet	feet	feet	feet	feet	feet	feet
11500	20.34	20.34	20.34	20.33	20.33	20.33	20.32	20.32	20.31	20.31
11520	20.31	20.3	20.3	20.3	20.29	20.29	20.29	20.28	20.28	20.28
11540	20.27	20.27	20.27	20.26	20.26	20.25	20.25	20.25	20.24	20.24
11560	20.24	20.23	20.23	20.23	20.22	20.22	20.22	20.21	20.21	20.21
11580	20.2	20.2	20.2	20.19	20.19	20.18	20.18	20.18	20.17	20.17
11600	20.17	20.16	20.16	20.16	20.15	20.15	20.15	20.14	20.14	20.14
11620	20.13	20.13	20.13	20.12	20.12	20.12	20.11	20.11	20.11	20.1
11640	20.1	20.09	20.09	20.09	20.08	20.08	20.08	20.07	20.07	20.07
11660	20.06	20.06	20.06	20.05	20.05	20.05	20.04	20.04	20.04	20.03
11680	20.03	20.03	20.02	20.02	20.02	20.01	20.01	20.01	20	20
11700	20	19.99	19.99	19.98	19.98	19.98	19.97	19.97	19.97	19.96
11720	19.96	19.96	19.95	19.95	19.95	19.94	19.94	19.94	19.93	19.93
11740	19.93	19.92	19.92	19.92	19.91	19.91	19.91	19.9	19.9	19.9
11760	19.89	19.89	19.89	19.88	19.88	19.88	19.87	19.87	19.87	19.86
11780	19.86	19.86	19.85	19.85	19.85	19.84	19.84	19.84	19.83	19.83
11800	19.83	19.82	19.82	19.82	19.81	19.81	19.81	19.8	19.8	19.8
11820	19.79	19.79	19.79	19.78	19.78	19.78	19.77	19.77	19.77	19.76
11840	19.76	19.76	19.75	19.75	19.75	19.74	19.74	19.74	19.73	19.73
11860	19.73	19.72	19.72	19.72	19.71	19.71	19.71	19.7	19.7	19.7
11880	19.69	19.69	19.69	19.68	19.68	19.68	19.67	19.67	19.67	19.66
11900	19.66	19.66	19.65	19.65	19.65	19.64	19.64	19.64	19.63	19.63
11920	19.63	19.62	19.62	19.62	19.61	19.61	19.61	19.6	19.6	19.6
11940	19.59	19.59	19.59	19.58	19.58	19.58	19.57	19.57	19.57	19.56
11960	19.56	19.56	19.55	19.55	19.55	19.54	19.54	19.54	19.53	19.53
11980	19.53	19.52	19.52	19.52	19.51	19.51	19.51	19.5	19.5	19.5
12000	19.5	19.49	19.49	19.49	19.48	19.48	19.48	19.47	19.47	19.47

TABLE 4-10
QUARTER WAVELENGTH VERTICAL
14000 - 14450 KILOHERTZ
USE: AMATEUR

Measurements, in feet, advance in 1 kHz increments from left to right across columns.

kHz	feet	feet	feet	feet	feet	feet	feet	feet	feet	feet
14000	16.71	16.71	16.7	16.7	16.7	16.7	16.69	16.69	16.69	16.69
14020	16.69	16.68	16.68	16.68	16.68	16.67	16.67	16.67	16.67	16.66
14040	16.66	16.66	16.66	16.65	16.65	16.65	16.65	16.65	16.64	16.64
14060	16.64	16.64	16.63	16.63	16.63	16.63	16.62	16.62	16.62	16.62
14080	16.61	16.61	16.61	16.61	16.6	16.6	16.6	16.6	16.6	16.59
14100	16.59	16.59	16.59	16.58	16.58	16.58	16.58	16.57	16.57	16.57
14120	16.57	16.56	16.56	16.56	16.56	16.56	16.55	16.55	16.55	16.55
14140	16.54	16.54	16.54	16.54	16.53	16.53	16.53	16.53	16.53	16.52
14160	16.52	16.52	16.52	16.51	16.51	16.51	16.51	16.5	16.5	16.5
14180	16.5	16.49	16.49	16.49	16.49	16.49	16.48	16.48	16.48	16.48
14200	16.47	16.47	16.47	16.47	16.46	16.46	16.46	16.46	16.46	16.45
14220	16.45	16.45	16.45	16.44	16.44	16.44	16.44	16.43	16.43	16.43
14240	16.43	16.43	16.42	16.42	16.42	16.42	16.41	16.41	16.41	16.41
14260	16.4	16.4	16.4	16.4	16.4	16.39	16.39	16.39	16.39	16.38
14280	16.38	16.38	16.38	16.37	16.37	16.37	16.37	16.37	16.36	16.36
14300	16.36	16.36	16.35	16.35	16.35	16.35	16.34	16.34	16.34	16.34
14320	16.34	16.33	16.33	16.33	16.33	16.32	16.32	16:32	16.32	16.32
14340	16.31	16.31	16.31	16.31	16.3	16.3	16.3	16.3	16.29	16.29
14360	16.29	16.29	16.29	16.28	16.28	16.28	16.28	16.27	16.27	16.27
14380	16.27	16.27	16.26	16.26	16.26	16.26	16.25	16.25	16.25	16.25
14400	16.25	16.24	16.24	16.24	16.24	16.23	16.23	16.23	16.23	16.22
14420	16.22	16.22	16.22	16.22	16.21	16.21	16.21	16.21	16.2	16.2
14440	16.2	16.2	16.2	16.19	16.19	16.19	16.19	16.18	16.18	16.18

TABLE 4-11
QUARTER WAVELENGTH VERTICAL
15500 - 16000 KILOHERTZ
USE: INT'L. BROADCAST

Measurements, in feet, advance in 1 kHz increments from left to right across columns.

kHz	feet	feet	feet	feet	feet	feet	feet	feet	feet	feet
15500	15.09	15.09	15.09	15.09	15.08	15.08	15.08	15.08	15.08	15.07
15520	15.07	15.07	15.07	15.07	15.06	15.06	15.06	15.06	15.06	15.05
15540	15.05	15.05	15.05	15.05	15.05	15.04	15.04	15.04	15.04	15.04
15560	15.03	15.03	15.03	15.03	15.03	15.02	15.02	15.02	15.02	15.02
15580	15.01	15.01	15.01	15.01	15.01	15	15	15	15	15
15600	15	14.99	14.99	14.99	14.99	14.99	14.98	14.98	14.98	14.98
15620	14.98	14.97	14.97	14.97	14.97	14.97	14.96	14.96	14.96	14.96
15640	14.96	14.95	14.95	14.95	14.95	14.95	14.95	14.94	14.94	14.94
15660	14.94	14.94	14.93	14.93	14.93	14.93	14.93	14.92	14.92	14.92
15680	14.92	14.92	14.91	14.91	14.91	14.91	14.91	14.91	14.9	14.9
15700	14.9	14.9	14.9	14.89	14.89	14.89	14.89	14.89	14.88	14.88
15720	14.88	14.88	14.88	14.87	14.87	14.87	14.87	14.87	14.87	14.86
15740	14.86	14.86	14.86	14.86	14.85	14.85	14.85	14.85	14.85	14.84
15760	14.84	14.84	14.84	14.84	14.84	14.83	14.83	14.83	14.83	14.83
15780	14.82	14.82	14.82	14.82	14.82	14.81	14.81	14.81	14.81	14.81
15800	14.81	14.8	14.8	14.8	14.8	14.8	14.79	14.79	14.79	14.79
15820	14.79	14.78	14.78	14.78	14.78	14.78	14.78	14.77	14.77	14.77
15840	14.77	14.77	14.76	14.76	14.76	14.76	14.76	14.75	14.75	14.75
15860	14.75	14.75	14.75	14.74	14.74	14.74	14.74	14.74	14.73	14.73
15880	14.73	14.73	14.73	14.72	14.72	14.72	14.72	14.72	14.72	14.71
15900	14.71	14.71	14.71	14.71	14.7	14.7	14.7	14.7	14.7	14.7
15920	14.69	14.69	14.69	14.69	14.69	14.68	14.68	14.68	14.68	14.68
15940	14.68	14.67	14.67	14.67	14.67	14.67	14.66	14.66	14.66	14.66
15960	14.66	14.65	14.65	14.65	14.65	14.65	14.65	14.64	14.64	14.64
15980	14.64	14.64	14.63	14.63	14.63	14.63	14.63	14.63	14.62	14.62
16000	14.62	14.62	14.62	14.61	14.61	14.61	14.61	14.61	14.61	14.6

TABLE 4-12
QUARTER WAVELENGTH VERTICAL
16500 - 17000 KILOHERTZ
USE: INT'L. BROADCAST

Measurements, in feet, advance in 1 kHz increments from left to right across columns.

kHz	feet	feet	feet	feet	feet	feet	feet	feet	feet	feet
16500	14.18	14.18	14.17	14.17	14.17	14.17	14.17	14.16	14.16	14.16
16520	14.16	14.16	14.16	14.15	14.15	14.15	14.15	14.15	14.15	14.14
16540	14.14	14.14	14.14	14.14	14.14	14.13	14.13	14.13	14.13	14.13
16560	14.13	14.12	14.12	14.12	14.12	14.12	14.12	14.11	14.11	14.11
16580	14.11	14.11	14.1	14.1	14.1	14.1	14.1	14.1	14.09	14.09
16600	14.09	14.09	14.09	14.09	14.08	14.08	14.08	14.08	14.08	14.08
16620	14.07	14.07	14.07	14.07	14.07	14.07	14.06	14.06	14.06	14.06
16640	14.06	14.06	14.05	14.05	14.05	14.05	14.05	14.05	14.04	14.04
16660	14.04	14.04	14.04	14.04	14.03	14.03	14.03	14.03	14.03	14.03
16680	14.02	14.02	14.02	14.02	14.02	14.02	14.01	14.01	14.01	14.01
16700	14.01	14.01	14	14	14	14	14	14	13.99	13.99
16720	13.99	13.99	13.99	13.99	13.98	13.98	13.98	13.98	13.98	13.98
16740	13.97	13.97	13.97	13.97	13.97	13.97	13.96	13.96	13.96	13.96
16760	13.96	13.96	13.95	13.95	13.95	13.95	13.95	13.95	13.94	13.94
16780	13.94	13.94	13.94	13.94	13.93	13.93	13.93	13.93	13.93	13.93
16800	13.92	13.92	13.92	13.92	13.92	13.92	13.91	13.91	13.91	13.91
16820	13.91	13.91	13.9	13.9	13.9	13.9	13.9	13.9	13.89	13.89
16840	13.89	13.89	13.89	13.89	13.88	13.88	13.88	13.88	13.88	13.88
16860	13.67	13.87	13.87	13.87	13.87	13.87	13.86	13.86	13.86	13.86
16880	13.86	13.86	13.85	13.85	13.85	13.85	13.85	13.85	13.84	13.84
16900	13.84	13.84	13.84	13.84	13.83	13.83	13.83	13.83	13.83	13.83
16920	13.82	13.82	13.82	13.82	13.82	13.82	13.81	13.81	13.81	13.81
16940	13.81	13.81	13.81	13.8	13.8	13.8	13.8	13.8	13.8	13.79
16960	13.79	13.79	13.79	13.79	13.79	13.78	13.78	13.78	13.78	13.78
16980	13.78	13.77	13.77	13.77	13.77	13.77	13.77	13.76	13.76	13.76
17000	13.76	13.76	13.76	13.75	13.75	13.75	13.75	13.75	13.75	13.75

TABLE 4-13
QUARTER WAVELENGTH VERTICAL
18500 - 19000 KILOHERTZ
USE: INT'L. BROADCAST

Measurements, in feet, advance in 1 kHz increments from left to right across columns.

kHz	feet	feet	feet	feet	feet	feet	feet	feet	feet	feet
18500	12.64	12.64	12.64	12.64	12.64	12.64	12.64	12.63	12.63	12.63
18520	12.63	12.63	12.63	12.63	12.62	12.62	12.62	12.62	12.62	12.62
18540	12.62	12.61	12.61	12.61	12.61	12.61	12.61	12.61	12.61	12.6
18560	12.6	12.6	12.6	12.6	12.6	12.6	12.59	12.59	12.59	12.59
18580	12.59	12.59	12.59	12.59	12.58	12.58	12.58	12.58	12.58	12.58
18600	12.58	12.57	12.57	12.57	12.57	12.57	12.57	12.57	12.56	12.56
18620	12.56	12.56	12.56	12.56	12.56	12.56	12.55	12.55	12.55	12.55
18640	12.55	12.55	12.55	12.54	12.54	12.54	12.54	12.54	12.54	12.54
18660	12.54	12.53	12.53	12.53	12.53	12.53	12.53	12.53	12.52	12.52
18680	12.52	12.52	12.52	12.52	12.52	12.52	12.51	12.51	12.51	12.51
18700	12.51	12.51	12.51	12.5	12.5	12.5	12.5	12.5	12.5	12.5
18720	12.5	12.49	12.49	12.49	12.49	12.49	12.49	12.49	12.48	12.48
18740	12.48	12.48	12.48	12.48	12.48	12.48	12.47	12.47	12.47	12.47
18760	12.47	12.47	12.47	12.46	12.46	12.46	12.46	12.46	12.46	12.46
18780	12.46	12.45	12.45	12.45	12.45	12.45	12.45	12.45	12.44	12.44
18800	12.44	12.44	12.44	12.44	12.44	12.44	12.43	12.43	12.43	12.43
18820	12.43	12.43	12.43	12.42	12.42	12.42	12.42	12.42	12.42	12.42
18840	12.42	12.41	12.41	12.41	12.41	12.41	12.41	12.41	12.4	12.4
18860	12.4	12.4	12.4	12.4	12.4	12.4	12.39	12.39	12.39	12.39
18880	12.39	12.39	12.39	12.39	12.38	12.38	12.38	12.38	12.38	12.38
18900	12.38	12.37	12.37	12.37	12.37	12.37	12.37	12.37	12.37	12.36
18920	12.36	12.36	12.36	12.36	12.36	12.36	12.36	12.35	12.35	12.35
18940	12.35	12.35	12.35	12.35	12.34	12.34	12.34	12.34	12.34	12.34
18960	12.34	12.34	12.33	12.33	12.33	12.33	12.33	12.33	12.33	12.33
18980	12.32	12.32	12.32	12.32	12.32	12.32	12.32	12.31	12.31	12.31
19000	12.31	12.31	12.31	12.31	12.31	12.3	12.3	12.3	12.3	12.3

TABLE 4-14
QUARTER WAVELENGTH VERTICAL
21000 - 21450 KILOHERTZ
USE: AMATEUR

Measurements, in feet, advance in 1 kHz increments from left to right across columns.

kHz	feet	feet	feet	feet	feet	feet	feet	feet	feet	feet
21000	11.14	11.14	11.14	11.13	11.13	11.13	11.13	11.13	11.13	11.13
21020	11.13	11.13	11.13	11.12	11.12	11.12	11.12	11.12	11.12	11.12
21040	11.12	11.12	11.11	11.11	11.11	11.11	11.11	11.11	11.11	11.11
21060	11.11	11.11	11.1	11.1	11.1	11.1	11.1	11.1	11.1	11.1
21080	11.1	11.09	11.09	11.09	11.09	11.09	11.09	11.09	11.09	11.09
21100	11.09	11.08	11.08	11.08	11.08	11.08	11.08	11.08	11.08	11.08
21120	11.07	11.07	11.07	11.07	11.07	11.07	11.07	11.07	11.07	11.07
21140	11.06	11.06	11.06	11.06	11.06	11.06	11.06	11.06	11.06	11.05
21160	11.05	11.05	11.05	11.05	11.05	11.05	11.05	11.05	11.05	11.04
21180	11.04	11.04	11.04	11.04	11.04	11.04	11.04	11.04	11.03	11.03
21200	11.03	11.03	11.03	11.03	11.03	11.03	11.03	11.03	11.02	11.02
21220	11.02	11.02	11.02	11.02	11.02	11.02	11.02	11.02	11.01	11.01
21240	11.01	11.01	11.01	11.01	11.01	11.01	11.01	11	11	11
21260	11	11	11	11	11	11	11	10.99	10.99	10.99
21280	10.99	10.99	10.99	10.99	10.99	10.99	10.99	10.98	10.98	10.98
21300	10.98	10.98	10.98	10.98	10.98	10.98	10.97	10.97	10.97	10.97
21320	10.97	10.97	10.97	10.97	10.97	10.97	10.96	10.96	10.96	10.96
21340	10.96	10.96	10.96	10.96	10.96	10.96	10.95	10.95	10.95	10.95
21360	10.95	10.95	10.95	10.95	10.95	10.94	10.94	10.94	10.94	10.94
21380	10.94	10.94	10.94	10.94	10.94	10.93	10.93	10.93	10.93	10.93
21400	10.93	10.93	10.93	10.93	10.93	10.92	10.92	10.92	10.92	10.92
21420	10.92	10.92	10.92	10.92	10.92	10.91	10.91	10.91	10.91	10.91
21440	10.91	10.91	10.91	10.91	10.91	10.9	10.9	10.9	10.9	10.9

TABLE 4-15
QUARTER WAVELENGTH VERTICAL
25500 - 26000 KILOHERTZ
USE: INT'L. BROADCAST

Measurements, in feet, advance in 1 kHz increments from left to right across columns.

kHz	feet	feet	feet	feet	feet	feet	feet	feet	feet	feet
25500	9.17	9.17	9.17	9.17	9.17	9.17	9.17	9.17	9.17	9.16
25520	9.16	9.16	9.16	9.16	9.16	9.16	9.16	9.16	9.16	9.16
25540	9.16	9.16	9.16	9.15	9.15	9.15	9.15	9.15	9.15	9.15
25560	9.15	9.15	9.15	9.15	9.15	9.15	9.15	9.14	9.14	9.14
25580	9.14	9.14	9.14	9.14	9.14	9.14	9.14	9.14	9.14	9.14
25600	9.14	9.13	9.13	9.13	9.13	9.13	9.13	9.13	9.13	9.13
25620	9.13	9.13	9.13	9.13	9.13	9.12	9.12	9.12	9.12	9.12
25640	9.12	9.12	9.12	9.12	9.12	9.12	9.12	9.12	9.12	9.11
25660	9.11	9.11	9.11	9.11	9.11	9.11	9.11	9.11	9.11	9.11
25680	9.11	9.11	9.11	9.11	9.1	9.1	9.1	9.1	9.1	9.1
25700	9.1	9.1	9.1	9.1	9.1	9.1	9.1	9.1	9.09	9.09
25720	9.09	9.09	9.09	9.09	9.09	9.09	9.09	9.09	9.09	9.09
25740	9.09	9.09	9.08	9.08	9.08	9.08	9.08	9.08	9.08	9.08
25760	9.08	9.08	9.08	9.08	9.08	9.08	9.07	9.07	9.07	9.07
25780	9.07	9.07	9.07	9.07	9.07	9.07	9.07	9.07	9.07	9.07
25800	9.06	9.06	9.06	9.06	9.06	9.06	9.06	9.06	9.06	9.06
25820	9.06	9.06	9.06	9.06	9.05	9.05	9.05	9.05	9.05	9.05
25840	9.05	9.05	9.05	9.05	9.05	9.05	9.05	9.05	9.05	9.04
25860	9.04	9.04	9.04	9.04	9.04	9.04	9.04	9.04	9.04	9.04
25880	9.04	9.04	9.04	9.03	9.03	9.03	9.03	9.03	9.03	9.03
25900	9.03	9.03	9.03	9.03	9.03	9.03	9.03	9.02	9.02	9.02
25920	9.02	9.02	9.02	9.02	9.02	9.02	9.02	9.02	9.02	9.02
25940	9.02	9.02	9.01	9.01	9.01	9.01	9.01	9.01	9.01	9.01
25960	9.01	9.01	9.01	9.01	9.01	9.01	9	9	9	9
25980	9	9	9	9	9	9	9	9	9	9
26000	9	8.99	8.99	8.99	8.99	8.99	8.99	8.99	8.99	8.99

TABLE 4-16
QUARTER WAVELENGTH VERTICAL
26500 - 27000 KILOHERTZ
USE: INT'L. BROADCAST

Measurements, in feet, advance in 1 kHz increments from left to right across columns.

kHz	feet	feet	feet	feet	feet	feet	feet	feet	feet	feet
26500	8.83	8.82	8.82	8.82	8.82	8.82	8.82	8.82	8.82	8.82
26520	8.82	8.82	8.82	8.82	8.82	8.82	8.81	8.81	8.81	8.81
26540	8.81	8.81	8.81	8.81	8.81	8.81	8.81	8.81	8.81	8.81
26560	8.81	8.8	8.8	8.8	8.8	8.8	8.8	8.8	8.8	8.8
26580	8.8	8.8	8.8	8.8	8.8	8.8	8.79	8.79	8.79	8.79
26600	8.79	8.79	8.79	8.79	8.79	8.79	8.79	8.79	8.79	8.79
26620	8.79	8.78	8.78	8.78	8.78	8.78	8.78	8.78	8.78	8.78
26640	8.78	8.78	8.78	8.78	8.78	8.78	8.77	8.77	8.77	8.77
26660	8.77	8.77	8.77	8.77	8.77	8.77	8.77	8.77	8.77	8.77
26680	8.77	8.76	8.76	8.76	8.76	8.76	8.76	8.76	8.76	8.76
26700	8.76	8.76	8.76	8.76	8.76	8.76	8.76	8.75	8.75	8.75
26720	8.75	8.75	8.75	8.75	8.75	8.75	8.75	8.75	8.75	8.75
26740	8.75	8.75	8.74	8.74	8.74	8.74	8.74	8.74	8.74	8.74
26760	8.74	8.74	8.74	8.74	8.74	8.74	8.74	8.73	8.73	8.73
26780	8.73	8.73	8.73	8.73	8.73	8.73	8.73	8.73	8.73	8.73
26800	8.73	8.73	8.73	8.72	8.72	8.72	8.72	8.72	8.72	8.72
26820	8.72	8.72	8.72	8.72	8.72	8.72	8.72	8.72	8.71	8.71
26840	8.71	8.71	8.71	8.71	8.71	8.71	8.71	8.71	8.71	8.71
26860	8.71	8.71	8.71	8.7	8.7	8.7	8.7	8.7	8.7	8.7
26880	8.7	8.7	8.7	8.7	8.7	8.7	8.7	8.7	8.7	8.69
26900	8.69	8.69	8.69	8.69	8.69	8.69	8.69	8.69	8.69	8.69
26920	8.69	8.69	8.69	8.69	8.68	8.68	8.68	8.68	8.68	8.68
26940	8.68	8.68	8.68	8.68	8.68	8.68	8.68	8.68	8.68	8.68
26960	8.67	8.67	8.67	8.67	8.67	8.67	8.67	8.67	8.67	8.67
26980	8.67	8.67	8.67	8.67	8.67	8.66	8.66	8.66	8.66	8.66
27000	8.66	8.66	8.66	8.66	8.66	8.66	8.66	8.66	8.66	8.66

TABLE 4-17
QUARTER WAVELENGTH VERTICAL
28500 - 29000 KILOHERTZ
USE: AMATEUR

Measurements, in feet, advance in 1 kHz increments from left to right across columns.

kHz	feet	feet	feet	feet	feet	feet	feet	feet	feet	feet
28500	8.21	8.2	8.2	8.2	8.2	8.2	8.2	8.2	8.2	8.2
28520	8.2	8.2	8.2	8.2	8.2	8.2	8.2	8.2	8.2	8.19
28540	8.19	8.19	8.19	8.19	8.19	8.19	8.19	8.19	8.19	8.19
28560	8.19	8.19	8.19	8.19	8.19	8.19	8.18	8.18	8.18	8.18
28580	8.18	8.18	8.18	8.18	8.18	8.18	8.18	8.18	8.18	8.18
28600	8.18	8.18	8.18	8.18	8.17	8.17	8.17	8.17	8.17	8.17
28620	8.17	8.17	8.17	8.17	8.17	8.17	8.17	8.17	8.17	8.17
28640	8.17	8.16	8.16	8.16	8.16	8.16	8.16	8.16	8.16	8.16
28660	8.16	8.16	8.16	8.16	8.16	8.16	8.16	8.16	8.16	8.15
28680	8.15	8.15	8.15	8.15	8.15	8.15	8.15	8.15	8.15	8.15
28700	8.15	8.15	8.15	8.15	8.15	8.15	8.14	8.14	8.14	8.14
28720	8.14	8.14	8.14	8.14	8.14	8.14	8.14	8.14	8.14	8.14
28740	8.14	8.14	8.14	8.14	8.13	8.13	8.13	8.13	8.13	8.13
28760	8.13	8.13	8.13	8.13	8.13	8.13	8.13	8.13	8.13	8.13
28780	8.13	8.13	8.12	8.12	8.12	8.12	8.12	8.12	8.12	8.12
28800	8.12	8.12	8.12	8.12	8.12	8.12	8.12	8.12	8.12	8.11
28820	8.11	8.11	8.11	8.11	8.11	8.11	8.11	8.11	8.11	8.11
28840	8.11	8.11	8.11	8.11	8.11	8.11	8.11	8.1	8.1	8.1
28860	8.1	8.1	8.1	8.1	8.1	8.1	8.1	8.1	8.1	8.1
28880	8.1	8.1	8.1	8.1	8.1	8.09	8.09	8.09	8.09	8.09
28900	8.09	8.09	8.09	8.09	8.09	8.09	8.09	8.09	8.09	8.09
28920	8.09	8.09	8.09	8.08	8.08	8.08	8.08	8.08	8.08	8.08
28940	8.08	8.08	8.08	8.08	8.08	8.08	8.08	8.08	8.08	8.08
28960	8.08	8.07	8.07	8.07	8.07	8.07	8.07	8.07	8.07	8.07
28980	8.07	8.07	8.07	8.07	8.07	8.07	8.07	8.07	8.07	8.06
29000	8.06	8.06	8.06	8.06	8.06	8.06	8.06	8.06	8.06	8.06

TABLE 4-18
1/4-WAVELENGTH VERTICAL
29000 - 29500 KILOHERTZ
USE: AMATEUR

Measurements, in feet, advance in 1 kHz increments from left to right across columns.

kHz	feet	feet	feet	feet	feet	feet	feet	feet	feet	feet
29000	8.06	8.06	8.06	8.06	8.06	8.06	8.06	8.06	8.06	8.06
29020	8.06	8.06	8.06	8.06	8.06	8.06	8.06	8.05	8.05	8.05
29040	8.05	8.05	8.05	8.05	8.05	8.05	8.05	8.05	8.05	8.05
29060	8.05	8.05	8.05	8.05	8.05	8.04	8.04	8.04	8.04	8.04
29080	8.04	8.04	8.04	8.04	8.04	8.04	8.04	8.04	8.04	8.04
29100	8.04	8.04	8.04	8.03	8.03	8.03	8.03	8.03	8.03	8.03
29120	8.03	8.03	8.03	8.03	8.03	8.03	8.03	8.03	8.03	8.03
29140	8.03	8.02	8.02	8.02	8.02	8.02	8.02	8.02	8.02	8.02
29160	8.02	8.02	8.02	8.02	8.02	8.02	8.02	8.02	8.02	8.01
29180	8.01	8.01	8.01	8.01	8.01	8.01	8.01	8.01	8.01	8.01
29200	8.01	8.01	8.01	8.01	8.01	8.01	8.01	8	8	8
29220	8	8	8	8	8	8	8	8	8	8
29240	8	8	8	8	8	8	7.99	7.99	7.99	7.99
29260	7.99	7.99	7.99	7.99	7.99	7.99	7.99	7.99	7.99	7.99
29280	7.99	7.99	7.99	7.99	7.98	7.98	7.98	7.98	7.98	7.98
29300	7.98	7.98	7.98	7.98	7.98	7.98	7.98	7.98	7.98	7.98
29320	7.98	7.98	7.97	7.97	7.97	7.97	7.97	7.97	7.97	7.97
29340	7.97	7.97	7.97	7.97	7.97	7.97	7.97	7.97	7.97	7.97
29360	7.97	7.96	7.96	7.96	7.96	7.96	7.96	7.96	7.96	7.96
29380	7.96	7.96	7.96	7.96	7.96	7.96	7.96	7.96	7.96	7.95
29400	7.95	7.95	7.95	7.95	7.95	7.95	7.95	7.95	7.95	7.95
29420	7.95	7.95	7.95	7.95	7.95	7.95	7.95	7.94	7.94	7.94
29440	7.94	7.94	7.94	7.94	7.94	7.94	7.94	7.94	7.94	7.94
29460	7.94	7.94	7.94	7.94	7.94	7.94	7.93	7.93	7.93	7.93
29480	7.93	7.93	7.93	7.93	7.93	7.93	7.93	7.93	7.93	7.93
29500	[illegible]	[illegible]	[illegible]	[illegible]	[illegible]	[illegible]	[illegible]	[illegible]	[illegible]	[illegible]

TABLE 4-19
QUARTER WAVELENGTH VERTICAL
29500 - 30000 KILOHERTZ
USE: AMATEUR

Measurements, in feet, advance in 1 kHz increments from left to right across columns.

kHz	feet	feet	feet	feet	feet	feet	feet	feet	feet	feet
29500	7.93	7.93	7.93	7.93	7.93	7.92	7.92	7.92	7.92	7.92
29520	7.92	7.92	7.92	7.92	7.92	7.92	7.92	7.92	7.92	7.92
29540	7.92	7.92	7.92	7.91	7.91	7.91	7.91	7.91	7.91	7.91
29560	7.91	7.91	7.91	7.91	7.91	7.91	7.91	7.91	7.91	7.91
29580	7.91	7.91	7.9	7.9	7.9	7.9	7.9	7.9	7.9	7.9
29600	7.9	7.9	7.9	7.9	7.9	7.9	7.9	7.9	7.9	7.9
29620	7.9	7.89	7.89	7.89	7.89	7.89	7.89	7.89	7.89	7.89
29640	7.89	7.89	7.89	7.89	7.89	7.89	7.89	7.89	7.89	7.88
29660	7.88	7.88	7.88	7.88	7.88	7.88	7.88	7.88	7.88	7.88
29680	7.88	7.88	7.88	7.88	7.88	7.88	7.88	7.88	7.87	7.87
29700	7.87	7.87	7.87	7.87	7.87	7.87	7.87	7.87	7.87	7.87
29720	7.87	7.87	7.87	7.87	7.87	7.87	7.87	7.86	7.86	7.86
29740	7.86	7.86	7.86	7.86	7.86	7.86	7.86	7.86	7.86	7.86
29760	7.86	7.86	7.86	7.86	7.86	7.86	7.85	7.85	7.85	7.85
29780	7.85	7.85	7.85	7.85	7.85	7.85	7.85	7.85	7.85	7.85
29800	7.85	7.85	7.85	7.85	7.85	7.84	7.84	7.84	7.84	7.84
29820	7.84	7.84	7.84	7.84	7.84	7.84	7.84	7.84	7.84	7.84
29840	7.84	7.84	7.84	7.84	7.83	7.83	7.83	7.83	7.83	7.83
29860	7.83	7.83	7.83	7.83	7.83	7.83	7.83	7.83	7.83	7.83
29880	7.83	7.83	7.83	7.82	7.82	7.82	7.82	7.82	7.82	7.82
29900	7.82	7.82	7.82	7.82	7.82	7.82	7.82	7.82	7.82	7.82
29920	7.82	7.82	7.81	7.81	7.81	7.81	7.81	7.81	7.81	7.81
29940	7.81	7.81	7.81	7.81	7.81	7.81	7.81	7.81	7.81	7.81
29960	7.81	7.8	7.8	7.8	7.8	7.8	7.8	7.8	7.8	7.8
29980	7.8	7.8	7.8	7.8	7.8	7.8	7.8	7.8	7.8	7.8
30000	7.8	7.79	7.79	7.79	7.79	7.79	7.79	7.79	7.79	7.79

Chapter 5

⅝-Wavelength Verticals

Figure 5-1 shows the ⅝-wavelength vertical antenna. Although the tables that follow give values for 80- and 40-meter operation, the majority of people find that this antenna is best suited for use at 20 meters and above. In fact, if an appropriate matching network is used, the same 20-foot radiator for a 10-meter ⅝-wavelength antenna is also useful on 15 meters and 20 meters (although not as a ⅝-wavelength.).

The impedance at the feedpoint is not easy to match with a simple piece of coaxial cable, so a matching network must be used. This antenna is often fed from the coaxial cable directly, but that always results in a high SWR. One of the matching systems given in a later chapter should be used.

The advantages of the ⅝-wavelength vertical include a low angel of radiation and a gain of approximately 3 dB over a ¼-wavelength vertical. The ⅝-wavelength vertical is especially popular among amateurs on 2 meters—most mobile 2-meter antennas are ⅝-wavelengths—and is popular among CBers.

The radials used when the ⅝-wavelength radiator is mounted above the ground must be ¼-wavelength. Use the dimensions given in Chapter 4 for these. The length of the radiator is given by:

$$L = 585/F$$

where L is the length in feet of the radiator and F is the frequency in megahertz (mHz).

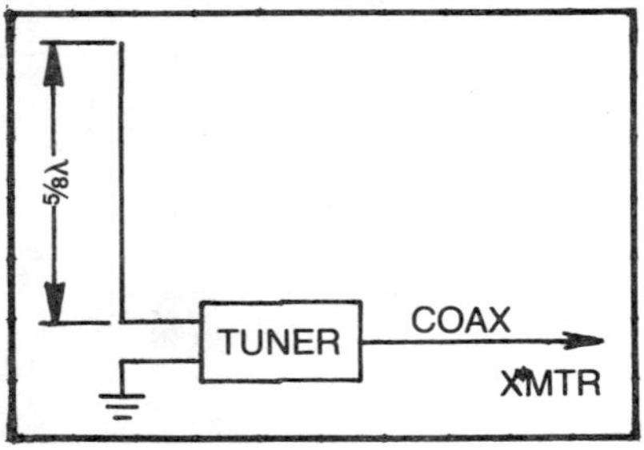

Fig. 5-1. ⅝-wavelength vertical.

TABLE 5-1
5/8-WAVELENGTH VERTICAL
300 - 500 KILOHERTZ
USE: MARINE/RADIO NAVIGATION

Measurements, in feet, advance in 1 kHz increments from left to right across columns.

kHz	feet	feet	feet	feet	feet	feet	feet	feet	feet	feet
300	1950	1943.52	1937.08	1930.69	1924.34	1918.03	1911.76	1905.53	1899.35	1893.2
310	1887.09	1881.02	1875	1869	1863.05	1857.14	1851.26	1845.42	1839.62	1833.85
320	1828.12	1822.42	1816.77	1811.14	1805.55	1800	1794.47	1788.99	1783.53	1778.11
330	1772.72	1767.37	1762.04	1756.75	1751.49	1746.26	1741.07	1735.9	1730.76	1725.66
340	1720.58	1715.54	1710.52	1705.53	1700.58	1695.65	1690.75	1685.87	1681.03	1676.21
350	1671.42	1666.66	1661.93	1657.22	1652.54	1647.88	1643.25	1638.65	1634.07	1629.52
360	1625	1620.49	1616.02	1611.57	1607.14	1602.73	1598.36	1594	1589.67	1585.36
370	1581.08	1576.81	1572.58	1568.36	1564.17	1560	1555.85	1551.72	1547.61	1543.53
380	1539.47	1535.43	1531.41	1527.41	1523.43	1519.48	1515.54	1511.62	1507.73	1503.85
390	1500	1496.16	1492.34	1488.54	1484.77	1481.01	1477.27	1473.55	1469.84	1466.16
400	1462.5	1458.85	1455.22	1451.61	1448.01	1444.44	1440.88	1437.34	1433.82	1430.31
410	1426.82	1423.35	1419.9	1416.46	1413.04	1409.63	1406.25	1402.87	1399.52	1396.18
420	1392.85	1389.54	1386.25	1382.97	1379.71	1376.47	1373.23	1370.02	1366.82	1363.63
430	1360.46	1357.3	1354.16	1351.03	1347.92	1344.82	1341.74	1338.67	1335.61	1332.57
440	1329.54	1326.53	1323.52	1320.54	1317.56	1314.6	1311.65	1308.72	1305.8	1302.89
450	1300	1297.11	1294.24	1291.39	1288.54	1285.71	1282.89	1280.08	1277.29	1274.5
460	1271.73	1268.98	1266.23	1263.49	1260.77	1258.06	1255.36	1252.67	1250	1247.33
470	1244.68	1242.03	1239.4	1236.78	1234.17	1231.57	1228.99	1226.41	1223.84	1221.29
480	1218.75	1216.21	1213.69	1211.18	1208.67	1206.18	1203.7	1201.23	1198.77	1196.31
490	1193.87	1191.44	1189.02	1186.61	1184.21	1181.81	1179.43	1177.06	1174.69	1172.34
500	1170	1167.66	1165.33	1163.02	1160.71	1158.41	1156.12	1153.84	1151.57	1149.31

TABLE 5-2
5/8-WAVELENGTH VERTICAL
540 - 1605 KILOHERTZ
USE: AM BROADCASTING

Measurements, in feet, advance in 1 kHz increments from left to right across columns.

kHz	feet	feet	feet	feet	feet	feet	feet	feet	feet	feet
540	1083.33	1073.39	1063.63	1054.05	1044.64	1035.39	1026.31	1017.39	1008.62	1000
590	991.52	983.19	975	966.94	959.01	951.21	943.54	936	928.57	921.25
640	914.06	906.97	900	893.12	886.36	879.69	873.13	866.66	860.29	854.01
690	847.82	841.72	835.71	829.78	823.94	818.18	812.5	806.89	801.36	795.91
740	790.54	785.23	780	774.83	769.73	764.7	759.74	754.83	750	745.22
790	740.5	735.84	731.25	726.7	722.22	717.79	713.41	709.09	704.81	700.59
840	696.42	692.3	688.23	684.21	680.23	676.3	672.41	668.57	664.77	661.01
890	657.3	653.63	650	646.4	642.85	639.34	635.86	632.43	629.03	625.66
940	622.34	619.04	615.78	612.56	609.37	606.21	603.09	600	596.93	593.9
990	590.9	587.93	585	582.08	579.2	576.35	573.52	570.73	567.96	565.21
1040	562.5	559.8	557.14	554.5	551.88	549.29	546.72	544.18	541.66	539.17
1090	536.69	534.24	531.81	529.41	527.02	524.66	522.32	520	517.69	515.41
1140	513.15	510.91	508.69	506.49	504.31	502.14	500	497.87	495.76	493.67
1190	491.59	489.53	487.5	485.47	483.47	481.48	479.5	477.55	475.6	473.68
1240	471.77	469.87	468	466.13	464.28	462.45	460.62	458.82	457.03	455.25
1290	453.48	451.73	450	448.27	446.56	444.86	443.18	441.5	439.84	438.2
1340	436.56	434.94	433.33	431.73	430.14	428.57	427	425.45	423.91	422.38
1370	420.86	419.35	417.85	416.37	414.89	413.42	411.97	410.52	409.09	407.66
1440	406.25	404.84	403.44	402.06	400.68	399.31	397.95	396.61	395.27	393.93
1490	392.61	391.3	390	388.7	387.41	386.13	384.86	383.6	382.35	381.1
1540	379.87	378.64	377.41	376.2	375	373.8	372.61	371.42	370.25	369.08
1590	367.92	366.77	365.62	364.48	363.35	362.22	361.11	360	358.89	357.79

TABLE 5-3
5/8-WAVELENGTH VERTICAL
1800 - 2000 KILOHERTZ
USE: AMATEUR

Measurements, in feet, advance in 1 kHz increments from left to right across columns.

kHz	feet	feet	feet	feet	feet	feet	feet	feet	feet	feet
1800	325	324.81	324.63	324.45	324.27	324.09	323.92	323.74	323.56	323.38
1810	323.2	323.02	322.84	322.66	322.49	322.31	322.13	321.95	321.78	321.6
1820	321.42	321.25	321.07	320.89	320.72	320.54	320.37	320.19	320.02	319.84
1830	319.67	319.49	319.32	319.14	318.97	318.8	318.62	318.45	318.28	318.1
1840	317.93	317.76	317.58	317.41	317.24	317.07	316.9	316.72	316.55	316.38
1850	316.21	316.04	315.87	315.7	315.53	315.36	315.19	315.02	314.85	314.68
1860	314.51	314.34	314.17	314	313.84	313.67	313.5	313.33	313.16	313
1870	312.83	312.66	312.5	312.33	312.16	312	311.83	311.66	311.5	311.33
1880	311.17	311	310.83	310.67	310.5	310.34	310.18	310.01	309.85	309.68
1890	309.52	309.36	309.19	309.03	308.87	308.7	308.54	308.38	308.21	308.05
1900	307.89	307.73	307.57	307.4	307.24	307.08	306.92	306.76	306.6	306.44
1910	306.28	306.12	305.96	305.8	305.64	305.48	305.32	305.16	305	304.84
1920	304.68	304.52	304.37	304.21	304.05	303.89	303.73	303.58	303.42	303.26
1930	303.1	302.95	302.79	302.63	302.48	302.32	302.16	302.01	301.85	301.7
1940	301.54	301.39	301.23	301.08	300.92	300.77	300.61	300.46	300.3	300.15
1950	300	299.84	299.69	299.53	299.38	299.23	299.07	298.92	298.77	298.62
1960	298.46	298.31	298.16	298.01	297.86	297.7	297.55	297.4	297.25	297.1
1970	296.95	296.8	296.65	296.5	296.35	296.2	296.05	295.9	295.75	295.6
1980	295.45	295.3	295.15	295	294.85	294.71	294.56	294.41	294.26	294.11
1990	293.96	293.82	293.67	293.52	293.38	293.23	293.08	292.93	292.79	292.64
2000	292.5	292.35	292.2	292.06	291.91	291.77	291.62	291.47	291.33	291.18

TABLE 5-4
5/8-WAVELENGTH VERTICAL
3500 - 4000 KILOHERTZ
USE: AMATEUR

Measurements, in feet, advance in 1 kHz increments from left to right across columns.

kHz	feet	feet	feet	feet	feet	feet	feet	feet	feet	feet
3500	167.14	167.04	166.95	166.85	166.76	166.66	166.57	166.47	166.38	166.28
3520	166.19	166.09	166	165.91	165.81	165.72	165.62	165.53	165.44	165.34
3540	165.25	165.16	165.06	164.97	164.88	164.78	164.69	164.6	164.51	164.41
3560	164.32	164.23	164.14	164.04	163.95	163.86	163.77	163.68	163.59	163.49
3580	163.4	163.31	163.22	163.13	163.04	162.95	162.86	162.77	162.68	162.59
3600	162.5	162.4	162.31	162.22	162.13	162.04	161.96	161.87	161.78	161.69
3620	161.6	161.51	161.42	161.33	161.24	161.15	161.06	160.97	160.89	160.8
3640	160.71	160.62	160.53	160.44	160.36	160.27	160.18	160.09	160.01	159.92
3660	159.83	159.74	159.66	159.57	159.48	159.4	159.31	159.22	159.14	159.05
3680	158.96	158.88	158.79	158.7	158.62	158.53	158.45	158.36	158.27	158.19
3700	158.1	158.02	157.93	157.85	157.76	157.68	157.59	157.51	157.42	157.34
3720	157.25	157.17	157.08	157	156.92	156.83	156.75	156.66	156.58	156.5
3740	156.41	156.33	156.25	156.16	156.08	156	155.91	155.83	155.75	155.66
3760	155.58	155.5	155.41	155.33	155.25	155.17	155.09	155	154.92	154.84
3780	154.76	154.68	154.59	154.51	154.43	154.35	154.27	154.19	154.1	154.02
3800	153.94	153.86	153.78	153.7	153.62	153.54	153.46	153.38	153.3	153.22
3820	153.14	153.06	152.98	152.9	152.82	152.74	152.66	152.58	152.5	152.42
3840	152.34	152.26	152.18	152.1	152.02	151.94	151.86	151.79	151.71	151.63
3860	151.55	151.47	151.39	151.31	151.24	151.16	151.08	151	150.92	150.85
3880	150.77	150.69	150.61	150.54	150.46	150.38	150.3	150.23	150.15	150.07
3900	150	149.92	149.84	149.76	149.69	149.61	149.53	149.46	149.38	149.31
3920	149.23	149.15	149.08	149	148.93	148.85	148.77	148.7	148.62	148.55
3940	148.47	148.4	148.32	148.25	148.17	148.1	148.02	147.95	147.87	147.8
3960	147.72	147.65	147.57	147.5	147.42	147.35	147.28	147.2	147.13	147.05
3980	146.98	146.91	146.83	146.76	146.69	146.61	146.54	146.46	146.39	146.32
4000	146.25	146.17	146.1	146.03	145.95	145.88	145.81	145.73	145.66	145.59

TABLE 5-5
5/8-WAVELENGTH VERTICAL
5500 - 6000 KILOHERTZ
USE: INT'L. BROADCAST

Measurements, in feet, advance in 1 kHz increments from left to right across columns.

kHz	feet	feet	feet	feet	feet	feet	feet	feet	feet	feet
5500	106.36	106.32	106.28	106.24	106.2	106.17	106.13	106.09	106.05	106.01
5520	105.97	105.93	105.9	105.86	105.82	105.78	105.74	105.71	105.67	105.63
5540	105.59	105.55	105.51	105.48	105.44	105.4	105.36	105.32	105.29	105.25
5560	105.21	105.17	105.14	105.1	105.06	105.02	104.98	104.95	104.91	104.87
5580	104.83	104.8	104.76	104.72	104.68	104.65	104.61	104.57	104.53	104.5
5600	104.46	104.42	104.38	104.35	104.31	104.27	104.24	104.2	104.16	104.12
5620	104.09	104.05	104.01	103.98	103.94	103.9	103.87	103.83	103.79	103.76
5640	103.72	103.68	103.64	103.61	103.57	103.53	103.5	103.46	103.42	103.39
5660	103.35	103.32	103.28	103.24	103.21	103.17	103.13	103.1	103.06	103.02
5680	102.99	102.95	102.92	102.88	102.84	102.81	102.77	102.73	102.7	102.66
5700	102.63	102.59	102.55	102.52	102.48	102.45	102.41	102.38	102.34	102.3
5720	102.27	102.23	102.2	102.16	102.12	102.09	102.05	102.02	101.98	101.95
5740	101.91	101.88	101.84	101.8	101.77	101.73	101.7	101.66	101.63	101.59
5760	101.56	101.52	101.49	101.45	101.42	101.38	101.35	101.31	101.28	101.24
5780	101.21	101.17	101.14	101.1	101.07	101.03	101	100.96	100.93	100.89
5800	100.86	100.82	100.79	100.75	100.72	100.68	100.65	100.61	100.58	100.55
5820	100.51	100.48	100.44	100.41	100.37	100.34	100.3	100.27	100.23	100.2
5840	100.17	100.13	100.1	100.06	100.03	100	99.76	99.93	99.89	99.86
5860	99.82	99.79	99.76	99.72	99.69	99.65	99.62	99.59	99.55	99.52
5880	99.48	99.45	99.42	99.38	99.35	99.32	99.28	99.25	99.21	99.18
5900	99.15	99.11	99.08	99.05	99.01	98.98	98.95	98.91	98.88	98.85
5920	98.81	98.78	98.75	98.71	98.68	98.65	98.61	98.58	98.55	98.51
5940	98.48	98.45	98.41	98.38	98.35	98.31	98.28	98.25	98.22	98.18
5960	98.15	98.12	98.08	98.05	98.02	97.98	97.95	97.92	97.89	97.85
5980	97.82	97.79	97.76	97.72	97.69	97.66	97.63	97.59	97.56	97.53
6000	97.5	97.46	97.43	97.4	97.37	97.33	97.3	97.27	97.24	97.2

TABLE 5-6
5/8-WAVELENGTH VERTICAL
6000 - 6500 KILOHERTZ
USE: INT'L. BROADCAST

Measurements, in feet, advance in 1 kHz increments from left to right across columns.

kHz	feet	feet	feet	feet	feet	feet	feet	feet	feet	feet
6000	97.5	97.46	97.43	97.4	97.37	97.33	97.3	97.27	97.24	97.2
6020	97.17	97.14	97.11	97.07	97.04	97.01	96.98	96.95	96.91	96.88
6040	96.85	96.82	96.79	96.75	96.72	96.69	96.66	96.63	96.59	96.56
6060	96.53	96.5	96.47	96.43	96.4	96.37	96.34	96.31	96.28	96.24
6080	96.21	96.18	96.15	96.12	96.09	96.05	96.02	95.99	95.96	95.93
6100	95.9	95.87	95.83	95.8	95.77	95.74	95.71	95.68	95.65	95.61
6120	95.58	95.55	95.52	95.49	95.46	95.43	95.4	95.37	95.33	95.3
6140	95.27	95.24	95.21	95.18	95.15	95.12	95.09	95.06	95.02	94.99
6160	94.96	94.93	94.9	94.87	94.84	94.81	94.78	94.75	94.72	94.69
6180	94.66	94.62	94.59	94.56	94.53	94.5	94.47	94.44	94.41	94.38
6200	94.35	94.32	94.29	94.26	94.23	94.2	94.17	94.14	94.11	94.08
6220	94.05	94.02	93.99	93.96	93.93	93.9	93.87	93.84	93.81	93.78
6240	93.75	93.71	93.68	93.65	93.62	93.6	93.57	93.54	93.51	93.48
6260	93.45	93.42	93.39	93.36	93.33	93.3	93.27	93.24	93.21	93.18
6280	93.15	93.12	93.09	93.06	93.03	93	92.97	92.94	92.91	92.88
6300	92.85	92.82	92.79	92.76	92.73	92.7	92.68	92.65	92.62	92.59
6320	92.56	92.53	92.5	92.47	92.44	92.41	92.38	92.35	92.32	92.3
6340	92.27	92.24	92.21	92.18	92.15	92.12	92.09	92.06	92.03	92.01
6360	91.98	91.95	91.92	91.89	91.86	91.83	91.8	91.77	91.75	91.72
6380	91.69	91.66	91.63	91.6	91.57	91.54	91.52	91.49	91.46	91.43
6400	91.4	91.37	91.34	91.32	91.29	91.26	91.23	91.2	91.17	91.14
6420	91.12	91.09	91.06	91.03	91	90.97	90.95	90.92	90.89	90.86
6440	90.83	90.81	90.78	90.75	90.72	90.69	90.66	90.64	90.61	90.58
6460	90.55	90.52	90.5	90.47	90.44	90.41	90.38	90.36	90.33	90.3
6480	90.27	90.24	90.22	90.19	90.16	90.13	90.11	90.08	90.05	90.02
6500	90	89.97	89.94	89.91	89.88	89.86	89.83	89.8	89.77	89.75

TABLE 5-7
5/8-WAVELENGTH VERTICAL
7000 - 7300 KILOHERTZ
USE: AMATEUR

Measurements, in feet, advance in 1 kHz increments from left to right across columns.

kHz	feet	feet	feet	feet	feet	feet	feet	feet	feet	feet
7000	83.57	83.54	83.52	83.49	83.47	83.45	83.42	83.4	83.38	83.35
7020	83.33	83.3	83.28	83.26	83.23	83.21	83.19	83.16	83.14	83.12
7040	83.09	83.07	83.04	83.02	83	82.97	82.95	82.93	82.9	82.88
7060	82.86	82.83	82.81	82.79	82.76	82.74	82.72	82.69	82.67	82.65
7080	82.62	82.6	82.58	82.55	82.53	82.51	82.48	82.46	82.44	82.41
7100	82.39	82.37	82.34	82.32	82.3	82.27	82.25	82.23	82.2	82.18
7120	82.16	82.13	82.11	82.09	82.07	82.04	82.02	82	81.97	81.95
7140	81.93	81.9	81.88	81.86	81.84	81.81	81.79	81.77	81.74	81.72
7160	81.7	81.68	81.65	81.63	81.61	81.58	81.56	81.54	81.52	81.49
7180	81.47	81.45	81.43	81.4	81.38	81.36	81.34	81.31	81.29	81.27
7200	81.25	81.22	81.2	81.18	81.15	81.13	81.11	81.09	81.06	81.04
7220	81.02	81	80.98	80.95	80.93	80.91	80.89	80.86	80.84	80.82
7240	80.8	80.77	80.75	80.73	80.71	80.68	80.66	80.64	80.62	80.6
7260	80.57	80.55	80.53	80.51	80.48	80.46	80.44	80.42	80.4	80.37
7280	80.35	80.33	80.31	80.29	80.26	80.24	80.22	80.2	80.18	80.15
7300	80.13	80.11	80.09	80.07	80.04	80.02	80	79.98	79.96	79.93

TABLE 5-8
5/8-WAVELENGTH VERTICAL
9500 - 10000 KILOHERTZ
USE: INT'L. BROADCAST

Measurements, in feet, advance in 1 kHz increments from left to right across columns.

kHz	feet	feet	feet	feet	feet	feet	feet	feet	feet	feet
9500	61.57	61.56	61.55	61.54	61.52	61.51	61.5	61.48	61.47	61.46
9520	61.44	61.43	61.42	61.41	61.39	61.38	61.37	61.35	61.34	61.33
9540	61.32	61.3	61.29	61.28	61.26	61.25	61.24	61.23	61.21	61.2
9560	61.19	61.17	61.16	61.15	61.14	61.12	61.11	61.1	61.09	61.07
9580	61.06	61.05	61.03	61.02	61.01	61	60.98	60.97	60.96	60.95
9600	60.93	60.92	60.91	60.89	60.88	60.87	60.86	60.84	60.83	60.82
9620	60.81	60.79	60.78	60.77	60.76	60.74	60.73	60.72	60.7	60.69
9640	60.68	60.67	60.65	60.64	60.63	60.62	60.6	60.59	60.58	60.57
9660	60.55	60.54	60.53	60.52	60.5	60.49	60.48	60.47	60.45	60.44
9680	60.43	60.42	60.4	60.39	60.38	60.37	60.35	60.34	60.33	60.32
9700	60.3	60.29	60.28	60.27	60.25	60.24	60.23	60.22	60.2	60.19
9720	60.18	60.17	60.16	60.14	60.13	60.12	60.11	60.09	60.08	60.07
9740	60.06	60.04	60.03	60.02	60.01	60	59.98	59.97	59.96	59.95
9760	59.93	59.92	59.91	59.9	59.88	59.87	59.86	59.85	59.84	59.82
9780	59.81	59.8	59.79	59.77	59.76	59.75	59.74	59.73	59.71	59.7
9800	59.69	59.68	59.66	59.65	59.64	59.63	59.62	59.6	59.59	59.58
9820	59.57	59.56	59.54	59.53	59.52	59.51	59.49	59.48	59.47	59.46
9840	59.45	59.43	59.42	59.41	59.4	59.39	59.37	59.36	59.35	59.34
9860	59.33	59.31	59.3	59.29	59.28	59.27	59.25	59.24	59.23	59.22
9880	59.21	59.19	59.18	59.17	59.16	59.15	59.13	59.12	59.11	59.1
9900	59.09	59.07	59.06	59.05	59.04	59.03	59.01	59	58.99	58.98
9920	58.97	58.95	58.94	58.93	58.92	58.91	58.9	58.88	58.87	58.86
9940	58.85	58.84	58.82	58.81	58.8	58.79	58.78	58.77	58.75	58.74
9960	58.73	58.72	58.71	58.69	58.68	58.67	58.66	58.65	58.64	58.62
9980	58.61	58.6	58.59	58.58	58.57	58.55	58.54	58.53	58.52	58.51
10000	58.5	58.48	58.47	58.46	58.45	58.44	58.42	58.41	58.4	58.39

TABLE 5-9
5/8-WAVELENGTH VERTICAL
11500 - 12000 KILOHERTZ
USE: INT'L. BROADCAST

Measurements, in feet, advance in 1 kHz increments from left to right across columns.

kHz	feet	feet	feet	feet	feet	feet	feet	feet	feet	feet
11500	50.86	50.86	50.85	50.84	50.83	50.82	50.81	50.8	50.79	50.79
11520	50.78	50.77	50.76	50.75	50.74	50.73	50.72	50.71	50.71	50.7
11540	50.69	50.68	50.67	50.66	50.65	50.64	50.64	50.63	50.62	50.61
11560	50.6	50.59	50.58	50.57	50.57	50.56	50.55	50.54	50.53	50.52
11580	50.51	50.5	50.5	50.49	50.48	50.47	50.46	50.45	50.44	50.43
11600	50.43	50.42	50.41	50.4	50.39	50.38	50.37	50.37	50.36	50.35
11620	50.34	50.33	50.32	50.31	50.3	50.3	50.29	50.28	50.27	50.26
11640	50.25	50.24	50.24	50.23	50.22	50.21	50.2	50.19	50.18	50.18
11660	50.17	50.16	50.15	50.14	50.13	50.12	50.11	50.11	50.1	50.09
11680	50.08	50.07	50.06	50.05	50.05	50.04	50.03	50.02	50.01	50
11700	50	49.99	49.98	49.97	49.96	49.95	49.94	49.94	49.93	49.92
11720	49.91	49.9	49.89	49.88	49.88	49.87	49.86	49.85	49.84	49.83
11740	49.82	49.82	49.81	49.8	49.79	49.78	49.77	49.77	49.76	49.75
11760	49.74	49.73	49.72	49.71	49.71	49.7	49.69	49.68	49.67	49.66
11780	49.66	49.65	49.64	49.63	49.62	49.61	49.6	49.6	49.59	49.58
11800	49.57	49.56	49.55	49.55	49.54	49.53	49.52	49.51	49.5	49.5
11820	49.49	49.48	49.47	49.46	49.45	49.45	49.44	49.43	49.42	49.41
11840	49.4	49.4	49.39	49.38	49.37	49.36	49.35	49.35	49.34	49.33
11860	49.32	49.31	49.3	49.3	49.29	49.28	49.27	49.26	49.25	49.25
11880	49.24	49.23	49.22	49.21	49.2	49.2	49.19	49.18	49.17	49.16
11900	49.15	49.15	49.14	49.13	49.12	49.11	49.11	49.1	49.09	49.08
11920	49.07	49.06	49.06	49.05	49.04	49.03	49.02	49.01	49.01	49
11940	48.99	48.98	48.97	48.97	48.96	48.95	48.94	48.93	48.92	48.92
11960	48.91	48.9	48.89	48.88	48.88	48.87	48.86	48.85	48.84	48.83
11980	48.83	48.82	48.81	48.8	48.79	48.79	48.78	48.77	48.76	48.75
12000	48.75	48.74	48.73	48.72	48.71	48.7	48.7	48.69	48.68	48.67

TABLE 5-10
5/8-WAVELENGTH VERTICAL
14000 - 14450 KILOHERTZ
USE: AMATEUR

Measurements, in feet, advance in 1 kHz increments from left to right across columns.

kHz	feet	feet	feet	feet	feet	feet	feet	feet	feet	feet
14000	41.78	41.77	41.77	41.76	41.76	41.75	41.74	41.74	41.73	41.73
14020	41.72	41.72	41.71	41.7	41.7	41.69	41.69	41.68	41.67	41.67
14040	41.66	41.66	41.65	41.64	41.64	41.63	41.63	41.62	41.61	41.61
14060	41.6	41.6	41.59	41.58	41.58	41.57	41.57	41.56	41.56	41.55
14080	41.54	41.54	41.53	41.53	41.52	41.51	41.51	41.5	41.5	41.49
14100	41.48	41.48	41.47	41.47	41.46	41.45	41.45	41.44	41.44	41.43
14120	41.43	41.42	41.41	41.41	41.4	41.4	41.39	41.38	41.38	41.37
14140	41.37	41.36	41.36	41.35	41.34	41.34	41.33	41.33	41.32	41.31
14160	41.31	41.3	41.3	41.29	41.29	41.28	41.27	41.27	41.26	41.26
14180	41.25	41.24	41.24	41.23	41.23	41.22	41.22	41.21	41.2	41.2
14200	41.19	41.19	41.18	41.17	41.17	41.16	41.16	41.15	41.15	41.14
14220	41.13	41.13	41.12	41.12	41.11	41.11	41.1	41.09	41.09	41.08
14240	41.08	41.07	41.06	41.06	41.05	41.05	41.04	41.04	41.03	41.02
14260	41.02	41.01	41.01	41	41	40.99	40.98	40.98	40.97	40.97
14280	40.96	40.96	40.95	40.94	40.94	40.93	40.93	40.92	40.92	40.91
14300	40.9	40.9	40.89	40.89	40.88	40.88	40.87	40.86	40.86	40.85
14320	40.85	40.84	40.84	40.83	40.82	40.82	40.81	40.81	40.8	40.8
14340	40.79	40.78	40.78	40.77	40.77	40.76	40.76	40.75	40.74	40.74
14360	40.73	40.73	40.72	40.72	40.71	40.7	40.7	40.69	40.69	40.68
14380	40.68	40.67	40.67	40.66	40.65	40.65	40.64	40.64	40.63	40.63
14400	40.62	40.61	40.61	40.6	40.6	40.59	40.59	40.58	40.57	40.57
14420	40.56	40.56	40.55	40.55	40.54	40.54	40.53	40.52	40.52	40.51
14440	40.51	40.5	40.5	40.49	40.49	40.48	40.47	40.47	40.46	40.46

TABLE 5-11
5/8-WAVELENGTH VERTICAL
15500 - 16000 KILOHERTZ
USE: INT'L. BROADCAST

Measurements, in feet, advance in 1 kHz increments from left to right across columns.

kHz	feet	feet	feet	feet	feet	feet	feet	feet	feet	feet
15500	37.74	37.73	37.73	37.72	37.72	37.71	37.71	37.7	37.7	37.69
15520	37.69	37.68	37.68	37.67	37.67	37.66	37.66	37.65	37.65	37.64
15540	37.64	37.63	37.63	37.63	37.62	37.62	37.61	37.61	37.6	37.6
15560	37.59	37.59	37.58	37.58	37.57	37.57	37.56	37.56	37.55	37.55
15580	37.54	37.54	37.53	37.53	37.52	37.52	37.51	37.51	37.5	37.5
15600	37.5	37.49	37.49	37.48	37.48	37.47	37.47	37.46	37.46	37.45
15620	37.45	37.44	37.44	37.43	37.43	37.42	37.42	37.41	37.41	37.4
15640	37.4	37.39	37.39	37.38	37.38	37.38	37.37	37.37	37.36	37.36
15660	37.35	37.35	37.34	37.34	37.33	37.33	37.32	37.32	37.31	37.31
15680	37.3	37.3	37.29	37.29	37.28	37.28	37.28	37.27	37.27	37.26
15700	37.26	37.25	37.25	37.24	37.24	37.23	37.23	37.22	37.22	37.21
15720	37.21	37.2	37.2	37.19	37.19	37.19	37.18	37.18	37.17	37.17
15740	37.16	37.16	37.15	37.15	37.14	37.14	37.13	37.13	37.12	37.12
15760	37.11	37.11	37.1	37.1	37.1	37.09	37.09	37.08	37.08	37.07
15780	37.07	37.06	37.06	37.05	37.05	37.04	37.04	37.03	37.03	37.03
15800	37.02	37.02	37.01	37.01	37	37	36.99	36.99	36.98	36.98
15820	36.97	36.97	36.96	36.96	36.95	36.95	36.95	36.94	36.94	36.93
15840	36.93	36.92	36.92	36.91	36.91	36.9	36.9	36.89	36.89	36.88
15860	36.88	36.88	36.87	36.87	36.86	36.86	36.85	36.85	36.84	36.84
15880	36.83	36.83	36.82	36.82	36.82	36.81	36.81	36.8	36.8	36.79
15900	36.79	36.78	36.78	36.77	36.77	36.76	36.76	36.76	36.75	36.75
15920	36.74	36.74	36.73	36.73	36.72	36.72	36.71	36.71	36.7	36.7
15940	36.7	36.69	36.69	36.68	36.68	36.67	36.67	36.66	36.66	36.65
15960	36.65	36.64	36.64	36.64	36.63	36.63	36.62	36.62	36.61	36.61
15980	36.6	36.6	36.59	36.59	36.58	36.58	36.58	36.57	36.57	36.56
16000	36.56	36.55	36.55	36.54	36.54	36.53	36.53	36.53	36.52	36.52

TABLE 5-12
5/8-WAVELENGTH VERTICAL
16500 - 17000 KILOHERTZ
USE: INT'L. BROADCAST

kHz	feet	feet	feet	feet	feet	feet	feet	feet	feet	feet
16500	35.45	35.45	35.44	35.44	35.43	35.43	35.42	35.42	35.42	35.41
16520	35.41	35.4	35.4	35.39	35.39	35.39	35.38	35.38	35.37	35.37
16540	35.36	35.36	35.36	35.35	35.35	35.34	35.34	35.33	35.33	35.33
16560	35.32	35.32	35.31	35.31	35.3	35.3	35.3	35.29	35.29	35.28
16580	35.28	35.27	35.27	35.27	35.26	35.26	35.25	35.25	35.24	35.24
16600	35.24	35.23	35.23	35.22	35.22	35.21	35.21	35.21	35.2	35.2
16620	35.19	35.19	35.19	35.18	35.18	35.17	35.17	35.16	35.16	35.16
16640	35.15	35.15	35.14	35.14	35.13	35.13	35.13	35.12	35.12	35.11
16660	35.11	35.1	35.1	35.1	35.09	35.09	35.08	35.08	35.08	35.07
16680	35.07	35.06	35.06	35.05	35.05	35.05	35.04	35.04	35.03	35.03
16700	35.02	35.02	35.02	35.01	35.01	35	35	35	34.99	34.99
16720	34.98	34.98	34.97	34.97	34.97	34.96	34.96	34.95	34.95	34.95
16740	34.94	34.94	34.93	34.93	34.92	34.92	34.92	34.91	34.91	34.9
16760	34.9	34.9	34.89	34.89	34.88	34.88	34.87	34.87	34.87	34.86
16780	34.86	34.85	34.85	34.85	34.84	34.84	34.83	34.83	34.82	34.82
16800	34.82	34.81	34.81	34.8	34.8	34.8	34.79	34.79	34.78	34.78
16820	34.78	34.77	34.77	34.76	34.76	34.75	34.75	34.75	34.74	34.74
16840	34.73	34.73	34.73	34.72	34.72	34.71	34.71	34.7	34.7	34.7
16860	34.69	34.69	34.68	34.68	34.68	34.67	34.67	34.66	34.66	34.66
16880	34.65	34.65	34.64	34.64	34.63	34.63	34.63	34.62	34.62	34.61
16900	34.61	34.61	34.6	34.6	34.59	34.59	34.59	34.58	34.58	34.57
16920	34.57	34.57	34.56	34.56	34.55	34.55	34.54	34.54	34.54	34.53
16940	34.53	34.52	34.52	34.52	34.51	34.51	34.5	34.5	34.5	34.49
16960	34.49	34.48	34.48	34.48	34.47	34.47	34.46	34.46	34.46	34.45
16980	34.45	34.44	34.44	34.44	34.43	34.43	34.42	34.42	34.41	34.41
17000	34.41	34.4	34.4	34.39	34.39	34.39	34.38	34.38	34.37	34.37

TABLE 5-13
5/8-WAVELENGTH VERTICAL
18500 - 19000 KILOHERTZ
USE: INT'L. BROADCAST

Measurements, in feet, advance in 1 kHz increments from left to right across columns.

kHz	feet	feet	feet	feet	feet	feet	feet	feet	feet	feet
18500	31.62	31.61	31.61	31.61	31.6	31.6	31.6	31.59	31.59	31.59
18520	31.58	31.58	31.58	31.57	31.57	31.57	31.56	31.56	31.56	31.55
18540	31.55	31.54	31.54	31.54	31.53	31.53	31.53	31.52	31.52	31.52
18560	31.51	31.51	31.51	31.5	31.5	31.5	31.49	31.49	31.49	31.48
18580	31.48	31.48	31.47	31.47	31.47	31.46	31.46	31.46	31.45	31.45
18600	31.45	31.44	31.44	31.44	31.43	31.43	31.43	31.42	31.42	31.42
18620	31.41	31.41	31.41	31.4	31.4	31.4	31.39	31.39	31.39	31.38
18640	31.38	31.38	31.37	31.37	31.37	31.36	31.36	31.36	31.35	31.35
18660	31.35	31.34	31.34	31.34	31.33	31.33	31.33	31.32	31.32	31.32
18680	31.31	31.31	31.31	31.3	31.3	31.3	31.29	31.29	31.29	31.28
18700	31.28	31.28	31.27	31.27	31.27	31.26	31.26	31.26	31.25	31.25
18720	31.25	31.24	31.24	31.23	31.23	31.23	31.22	31.22	31.22	31.21
18740	31.21	31.21	31.2	31.2	31.2	31.2	31.19	31.19	31.19	31.18
18760	31.18	31.18	31.17	31.17	31.17	31.16	31.16	31.16	31.15	31.15
18780	31.15	31.14	31.14	31.14	31.13	31.13	31.13	31.12	31.12	31.12
18800	31.11	31.11	31.11	31.1	31.1	31.1	31.09	31.09	31.09	31.08
18820	31.08	31.08	31.07	31.07	31.07	31.06	31.06	31.06	31.05	31.05
18840	31.05	31.04	31.04	31.04	31.03	31.03	31.03	31.02	31.02	31.02
18860	31.01	31.01	31.01	31	31	31	30.99	30.99	30.99	30.98
18880	30.98	30.98	30.97	30.97	30.97	30.96	30.96	30.96	30.95	30.95
18900	30.95	30.94	30.94	30.94	30.93	30.93	30.93	30.92	30.92	30.92
18920	30.91	30.91	30.91	30.9	30.9	30.9	30.9	30.89	30.89	30.89
18940	30.88	30.88	30.88	30.87	30.87	30.87	30.86	30.86	30.86	30.85
18960	30.85	30.85	30.84	30.84	30.84	30.83	30.83	30.83	30.82	30.82
18980	30.82	30.81	30.81	30.81	30.8	30.8	30.8	30.79	30.79	30.79
19000	30.78	30.78	30.78	30.77	30.77	30.77	30.77	30.76	30.76	30.76

TABLE 5-14
5/8-WAVELENGTH VERTICAL
21000 - 21450 KILOHERTZ
USE: AMATEUR

Measurements, in feet, advance in 1 kHz increments from left to right across columns.

kHz	feet	feet	feet	feet	feet	feet	feet	feet	feet	feet
21000	27.85	27.85	27.85	27.84	27.84	27.84	27.84	27.83	27.83	27.83
21020	27.83	27.82	27.82	27.82	27.82	27.81	27.81	27.81	27.8	27.8
21040	27.8	27.8	27.79	27.79	27.79	27.79	27.78	27.78	27.78	27.78
21060	27.77	27.77	27.77	27.76	27.76	27.76	27.76	27.75	27.75	27.75
21080	27.75	27.74	27.74	27.74	27.74	27.73	27.73	27.73	27.73	27.72
21100	27.72	27.72	27.71	27.71	27.71	27.71	27.7	27.7	27.7	27.7
21120	27.69	27.69	27.69	27.69	27.68	27.68	27.68	27.68	27.67	27.67
21140	27.67	27.67	27.66	27.66	27.66	27.65	27.65	27.65	27.65	27.64
21160	27.64	27.64	27.64	27.63	27.63	27.63	27.63	27.62	27.62	27.62
21180	27.62	27.61	27.61	27.61	27.6	27.6	27.6	27.6	27.59	27.59
21200	27.59	27.59	27.58	27.58	27.58	27.58	27.57	27.57	27.57	27.57
21220	27.56	27.56	27.56	27.56	27.55	27.55	27.55	27.55	27.54	27.54
21240	27.54	27.53	27.53	27.53	27.53	27.52	27.52	27.52	27.52	27.51
21260	27.51	27.51	27.51	27.5	27.5	27.5	27.5	27.49	27.49	27.49
21280	27.49	27.48	27.48	27.48	27.48	27.47	27.47	27.47	27.46	27.46
21300	27.46	27.46	27.45	27.45	27.45	27.45	27.44	27.44	27.44	27.44
21320	27.43	27.43	27.43	27.43	27.42	27.42	27.42	27.42	27.41	27.41
21340	27.41	27.41	27.4	27.4	27.4	27.4	27.39	27.39	27.39	27.39
21360	27.38	27.38	27.38	27.37	27.37	27.37	27.37	27.36	27.36	27.36
21380	27.36	27.35	27.35	27.35	27.35	27.34	27.34	27.34	27.34	27.33
21400	27.33	27.33	27.33	27.32	27.32	27.32	27.32	27.31	27.31	27.31
21420	27.31	27.3	27.3	27.3	27.3	27.29	27.29	27.29	27.29	27.28
21440	27.28	27.28	27.28	27.27	27.27	27.27	27.27	27.26	27.26	27.26

TABLE 5-15
5/8-WAVELENGTH VERTICAL
25500 - 26000 KILOHERTZ
USE: INT'L. BROADCAST

Measurements, in feet, advance in 1 kHz increments from left to right across columns.

kHz	feet	feet	feet	feet	feet	feet	feet	feet	feet	feet
25500	22.94	22.93	22.93	22.93	22.93	22.93	22.93	22.92	22.92	22.92
25520	22.92	22.92	22.91	22.91	22.91	22.91	22.91	22.91	22.9	22.9
25540	22.9	22.9	22.9	22.89	22.89	22.89	22.89	22.89	22.89	22.88
25560	22.88	22.88	22.88	22.88	22.88	22.87	22.87	22.87	22.87	22.87
25580	22.86	22.86	22.86	22.86	22.86	22.86	22.85	22.85	22.85	22.85
25600	22.85	22.84	22.84	22.84	22.84	22.84	22.84	22.83	22.83	22.83
25620	22.83	22.83	22.83	22.82	22.82	22.82	22.82	22.82	22.81	22.81
25640	22.81	22.81	22.81	22.81	22.8	22.8	22.8	22.8	22.8	22.79
25660	22.79	22.79	22.79	22.79	22.79	22.78	22.78	22.78	22.78	22.78
25680	22.78	22.77	22.77	22.77	22.77	22.77	22.76	22.76	22.76	22.76
25700	22.76	22.76	22.75	22.75	22.75	22.75	22.75	22.75	22.74	22.74
25720	22.74	22.74	22.74	22.73	22.73	22.73	22.73	22.73	22.73	22.72
25740	22.72	22.72	22.72	22.72	22.72	22.71	22.71	22.71	22.71	22.71
25760	22.7	22.7	22.7	22.7	22.7	22.7	22.69	22.69	22.69	22.69
25780	22.69	22.69	22.68	22.68	22.68	22.68	22.68	22.67	22.67	22.67
25800	22.67	22.67	22.67	22.66	22.66	22.66	22.66	22.66	22.66	22.65
25820	22.65	22.65	22.65	22.65	22.64	22.64	22.64	22.64	22.64	22.64
25840	22.63	22.63	22.63	22.63	22.63	22.63	22.62	22.62	22.62	22.62
25860	22.62	22.62	22.61	22.61	22.61	22.61	22.61	22.6	22.6	22.6
25880	22.6	22.6	22.6	22.59	22.59	22.59	22.59	22.59	22.59	22.58
25900	22.58	22.58	22.58	22.58	22.57	22.57	22.57	22.57	22.57	22.57
25920	22.56	22.56	22.56	22.56	22.56	22.56	22.55	22.55	22.55	22.55
25940	22.55	22.55	22.54	22.54	22.54	22.54	22.54	22.53	22.53	22.53
25960	22.53	22.53	22.53	22.52	22.52	22.52	22.52	22.52	22.52	22.51
25980	22.51	22.51	22.51	22.51	22.51	22.5	22.5	22.5	22.5	22.5
26000	22.5	22.49	22.49	22.49	22.49	22.49	22.48	22.48	22.48	22.48

TABLE 5-16
5/8-WAVELENGTH VERTICAL
26500 - 27000 KILOHERTZ
USE: INT'L. BROADCAST

Measurements, in feet, advance in 1 kHz increments from left to right across columns.

kHz	feet	feet	feet	feet	feet	feet	feet	feet	feet	feet
26500	22.07	22.07	22.07	22.07	22.06	22.06	22.06	22.06	22.06	22.06
26520	22.05	22.05	22.05	22.05	22.05	22.05	22.04	22.04	22.04	22.04
26540	22.04	22.04	22.03	22.03	22.03	22.03	22.03	22.03	22.02	22.02
26560	22.02	22.02	22.02	22.02	22.01	22.01	22.01	22.01	22.01	22.01
26580	22	22	22	22	22	22	21.99	21.99	21.99	21.99
26600	21.99	21.99	21.98	21.98	21.98	21.98	21.98	21.98	21.97	21.97
26620	21.97	21.97	21.97	21.97	21.96	21.96	21.96	21.96	21.96	21.96
26640	21.95	21.95	21.95	21.95	21.95	21.95	21.94	21.94	21.94	21.94
26660	21.94	21.94	21.93	21.93	21.93	21.93	21.93	21.93	21.92	21.92
26680	21.92	21.92	21.92	21.92	21.91	21.91	21.91	21.91	21.	21.91
26700	21.91	21.9	21.9	21.9	21.9	21.9	21.9	21.	21.89	21.89
26720	21.89	21.89	21.89	21.88	21.88	21.88	21.88	21.88	21.88	21.87
26740	21.87	21.87	21.87	21.87	21.87	21.86	21.86	21.86	21.86	21.86
26760	21.86	21.85	21.85	21.85	21.85	21.85	21.85	21.84	21.84	21.84
26780	21.84	21.84	21.84	21.83	21.83	21.83	21.83	21.83	21.83	21.82
26800	21.82	21.82	21.82	21.82	21.82	21.82	21.81	21.81	21.81	21.81
26820	21.81	21.81	21.8	21.8	21.8	21.8	21.8	21.8	21.79	21.79
26840	21.79	21.79	21.79	21.79	21.78	21.78	21.78	21.78	21.78	21.78
26860	21.77	21.77	21.77	21.77	21.77	21.77	21.76	21.76	21.76	21.76
26880	21.76	21.76	21.76	21.75	21.75	21.75	21.75	21.75	21.75	21.74
26900	21.74	21.74	21.74	21.74	21.74	21.73	21.73	21.73	21.73	21.73
26920	21.73	21.72	21.72	21.72	21.72	21.72	21.72	21.71	21.71	21.71
26940	21.71	21.71	21.71	21.71	21.7	21.7	21.7	21.7	21.7	21.7
26960	21.69	21.69	21.69	21.69	21.69	21.69	21.68	21.68	21.68	21.68
26980	21.68	21.68	21.67	21.67	21.67	21.67	21.67	21.67	21.66	21.66
27000	21.66	21.66	21.66	21.66	21.66	21.65	21.65	21.65	21.65	[illegible]

TABLE 5-17
5/8-WAVELENGTH VERTICAL
28500 - 29000 KILOHERTZ
USE: AMATEUR

Measurements, in feet, advance in 1 kHz increments from left to right across columns.

kHz	feet	feet	feet	feet	feet	feet	feet	feet	feet	feet
28500	20.52	20.52	20.52	20.52	20.52	20.51	20.51	20.51	20.51	20.51
28520	20.51	20.51	20.5	20.5	20.5	20.5	20.5	20.5	20.5	20.49
28540	20.49	20.49	20.49	20.49	20.49	20.49	20.48	20.48	20.48	20.48
28560	20.48	20.48	20.48	20.47	20.47	20.47	20.47	20.47	20.47	20.47
28580	20.46	20.46	20.46	20.46	20.46	20.46	20.46	20.45	20.45	20.45
28600	20.45	20.45	20.45	20.45	20.44	20.44	20.44	20.44	20.44	20.44
28620	20.44	20.43	20.43	20.43	20.43	20.43	20.43	20.43	20.42	20.42
28640	20.42	20.42	20.42	20.42	20.42	20.41	20.41	20.41	20.41	20.41
28660	20.41	20.41	20.4	20.4	20.4	20.4	20.4	20.4	20.4	20.39
28680	20.39	20.39	20.39	20.39	20.39	20.39	20.38	20.38	20.38	20.38
28700	20.38	20.38	20.38	20.37	20.37	20.37	20.37	20.37	20.37	20.37
28720	20.36	20.36	20.36	20.36	20.36	20.36	20.36	20.35	20.35	20.35
28740	20.35	20.35	20.35	20.35	20.34	20.34	20.34	20.34	20.34	20.34
28760	20.34	20.33	20.33	20.33	20.33	20.33	20.33	20.33	20.32	20.32
28780	20.32	20.32	20.32	20.32	20.32	20.31	20.31	20.31	20.31	20.31
28800	20.31	20.31	20.3	20.3	20.3	20.3	20.3	20.3	20.3	20.29
28820	20.29	20.29	20.29	20.29	20.29	20.29	20.28	20.28	20.28	20.28
28840	20.28	20.28	20.28	20.28	20.27	20.27	20.27	20.27	20.27	20.27
28860	20.27	20.26	20.26	20.26	20.26	20.26	20.26	20.26	20.25	20.25
28880	20.25	20.25	20.25	20.25	20.25	20.24	20.24	20.24	20.24	20.24
28900	20.24	20.24	20.23	20.23	20.23	20.23	20.23	20.23	20.23	20.22
28920	20.22	20.22	20.22	20.22	20.22	20.22	20.21	20.21	20.21	20.21
28940	20.21	20.21	20.21	20.21	20.2	20.2	20.2	20.2	20.2	20.2
28960	20.2	20.19	20.19	20.19	20.19	20.19	20.19	20.19	20.18	20.18
28980	20.18	20.18	20.18	20.18	20.18	20.17	20.17	20.17	20.17	20.17
29000	20.17	20.17	20.16	20.16	20.16	20.16	20.16	20.16	20.16	20.15

TABLE 5-18
5/8-WAVELENGTH VERTICAL
29000 - 29500 KILOHERTZ
USE: AMATEUR

Measurements, in feet, advance in 1 kHz increments from left to right across columns.

kHz	feet	feet	feet	feet	feet	feet	feet	feet	feet	feet
29000	20.17	20.17	20.16	20.16	20.16	20.16	20.16	20.16	20.16	20.15
29020	20.15	20.15	20.15	20.15	20.15	20.15	20.15	20.14	20.14	20.14
29040	20.14	20.14	20.14	20.14	20.13	20.13	20.13	20.13	20.13	20.13
29060	20.13	20.12	20.12	20.12	20.12	20.12	20.12	20.12	20.11	20.11
29080	20.11	20.11	20.11	20.11	20.11	20.11	20.1	20.1	20.1	20.1
29100	20.1	20.1	20.1	20.09	20.09	20.09	20.09	20.09	20.09	20.09
29120	20.08	20.08	20.08	20.08	20.08	20.08	20.08	20.07	20.07	20.07
29140	20.07	20.07	20.07	20.07	20.06	20.06	20.06	20.06	20.06	20.06
29160	20.06	20.06	20.05	20.05	20.05	20.05	20.05	20.05	20.05	20.04
29180	20.04	20.04	20.04	20.04	20.04	20.04	20.03	20.03	20.03	20.03
29200	20.03	20.03	20.03	20.03	20.02	20.02	20.02	20.02	20.02	20.02
29220	20.02	20.01	20.01	20.01	20.01	20.01	20.01	20.01	20	20
29240	20	20	20	20	20	20	19.99	19.99	19.99	19.99
29260	19.99	19.99	19.99	19.98	19.98	19.98	19.98	19.98	19.98	19.98
29280	19.97	19.97	19.97	19.97	19.97	19.97	19.97	19.96	19.96	19.96
29300	19.96	19.96	19.96	19.96	19.96	19.95	19.95	19.95	19.95	19.95
29320	19.95	19.95	19.94	19.94	19.94	19.94	19.94	19.94	19.94	19.94
29340	19.93	19.93	19.93	19.93	19.93	19.93	19.93	19.92	19.92	19.92
29360	19.92	19.92	19.92	19.92	19.91	19.91	19.91	19.91	19.91	19.91
29380	19.91	19.91	19.9	19.9	19.9	19.9	19.9	19.9	19.9	19.89
29400	19.89	19.89	19.89	19.89	19.89	19.89	19.88	19.88	19.88	19.88
29420	19.88	19.88	19.88	19.88	19.87	19.87	19.87	19.87	19.87	19.87
29440	19.87	19.86	19.86	19.86	19.86	19.86	19.86	19.86	19.86	19.85
29460	19.85	19.85	19.85	19.85	19.85	19.85	19.84	19.84	19.84	19.84
29480	19.84	19.84	19.84	19.83	19.83	19.83	19.83	19.83	19.83	19.83
29500	19.83	19.82	19.82	19.82	19.82	19.82	19.82	19.82	19.81	19.81

TABLE 5-19
5/8-WAVELENGTH VERTICAL
29500 - 30000 KILOHERTZ
USE: AMATEUR

Measurements, in feet, advance in 1 kHz increments from left to right across columns.

kHz	feet	feet	feet	feet	feet	feet	feet	feet	feet	feet
29500	19.83	19.82	19.82	19.82	19.82	19.82	19.82	19.82	19.81	19.81
29520	19.81	19.81	19.81	19.81	19.81	19.81	19.8	19.8	19.8	19.8
29540	19.8	19.8	19.8	19.79	19.79	19.79	19.79	19.79	19.79	19.79
29560	19.79	19.78	19.78	19.78	19.78	19.78	19.78	19.78	19.77	19.77
29580	19.77	19.77	19.77	19.77	19.77	19.77	19.76	19.76	19.76	19.76
29600	19.76	19.76	19.76	19.75	19.75	19.75	19.75	19.75	19.75	19.75
29620	19.75	19.74	19.74	19.74	19.74	19.74	19.74	19.74	19.73	19.73
29640	19.73	19.73	19.73	19.73	19.73	19.73	19.72	19.72	19.72	19.72
29660	19.72	19.72	19.72	19.71	19.71	19.71	19.71	19.71	19.71	19.71
29680	19.71	19.7	19.7	19.7	19.7	19.7	19.7	19.7	19.69	19.69
29700	19.69	19.69	19.69	19.69	19.69	19.69	19.68	19.68	19.68	19.68
29720	19.68	19.68	19.68	19.67	19.67	19.67	19.67	19.67	19.67	19.67
29740	19.67	19.66	19.66	19.66	19.66	19.66	19.66	19.66	19.65	19.65
29760	19.65	19.65	19.65	19.65	19.65	19.65	19.64	19.64	19.64	19.64
29780	19.64	19.64	19.64	19.64	19.63	19.63	19.63	19.63	19.63	19.63
29800	19.63	19.62	19.62	19.62	19.62	19.62	19.62	19.62	19.62	19.61
29820	19.61	19.61	19.61	19.61	19.61	19.61	19.6	19.6	19.6	19.6
29840	19.6	19.6	19.6	19.6	19.59	19.59	19.59	19.59	19.59	19.59
29860	19.59	19.59	19.58	19.58	19.58	19.58	19.58	19.58	19.58	19.57
29880	19.57	19.57	19.57	19.57	19.57	19.57	19.57	19.56	19.56	19.56
29900	19.56	19.56	19.56	19.56	19.55	19.55	19.55	19.55	19.55	19.55
29920	19.55	19.55	19.54	19.54	19.54	19.54	19.54	19.54	19.54	19.54
29940	19.53	19.53	19.53	19.53	19.53	19.53	19.53	19.52	19.52	19.52
29960	19.52	19.52	19.52	19.52	19.52	19.51	19.51	19.51	19.51	19.51
29980	19.51	19.51	19.51	19.5	19.5	19.5	19.5	19.5	19.5	19.5
30000	19.5	19.49	19.49	19.49	19.49	19.49	19.49	19.49	19.48	19.48

Chapter 6
¾-Wavelength Verticals

A ¾-wavelength vertical can be built as a compromise for the ⅝-wavelength antenna. Although some deterioration of performance is noted, the impedance at the feedpoint is lower than for the ⅝-wavelength case, making it easier to feed the antenna without a tuner or matching device. Again, the radials used when the antenna is mounted above the ground, must be quarter wavelength. The length of the radiator is found from

$$L = \frac{702}{F}$$

where L is the length in feet and F is the frequency in megahertz (MHz).

Tables giving the lengths appropriate for several popular frequencies are given in the following tables.

TABLE 6-1
3/4-WAVELENGTH VERTICAL
300 - 500 KILOHERTZ
USE: MARINE/RADIO NAVIGATION

Measurements, in feet, advance in 1 kHz increments from left to right across columns.

kHz	feet	feet	feet	feet	feet	feet	feet	feet	feet	feet
300	2340	2332.22	2324.5	2316.83	2309.21	2301.63	2294.11	2286.64	2279.22	2271.84
310	2264.51	2257.23	2250	2242.81	2235.66	2228.57	2221.51	2214.51	2207.54	2200.62
320	2193.75	2186.91	2180.12	2173.37	2166.66	2160	2153.37	2146.78	2140.24	2133.73
330	2127.27	2120.84	2114.45	2108.1	2101.79	2095.52	2089.28	2083.08	2076.92	2070.79
340	2064.7	2058.65	2052.63	2046.64	2040.69	2034.78	2028.9	2023.05	2017.24	2011.46
350	2005.71	2000	1994.31	1988.66	1983.05	1977.46	1971.91	1966.38	1960.89	1955.43
360	1950	1944.59	1939.22	1933.88	1928.57	1923.28	1918.03	1912.8	1907.6	1902.43
370	1897.29	1892.18	1887.09	1882.03	1877	1872	1867.02	1862.06	1857.14	1852.24
380	1847.36	1842.51	1837.69	1832.89	1828.12	1823.37	1818.65	1813.95	1809.27	1804.62
390	1800	1795.39	1790.81	1786.25	1781.72	1777.21	1772.72	1768.26	1763.81	1759.39
400	1755	1750.62	1746.26	1741.93	1737.62	1733.33	1729.06	1724.81	1720.58	1716.38
410	1712.19	1708.02	1703.88	1699.75	1695.65	1691.56	1687.5	1683.45	1679.42	1675.41
420	1671.42	1667.45	1663.5	1659.57	1655.66	1651.76	1647.88	1644.02	1640.18	1636.36
430	1632.55	1628.77	1625	1621.24	1617.51	1613.79	1610.09	1606.4	1602.73	1599.08
440	1595.45	1591.83	1588.23	1584.65	1581.08	1577.52	1573.99	1570.46	1566.96	1563.47
450	1560	1556.54	1553.09	1549.66	1546.25	1542.85	1539.47	1536.1	1532.75	1529.41
460	1526.08	1522.77	1519.48	1516.19	1512.93	1509.67	1506.43	1503.21	1500	1496.8
470	1493.61	1490.44	1487.28	1484.14	1481.01	1477.89	1474.78	1471.69	1468.61	1465.55
480	1462.5	1459.45	1456.43	1453.41	1450.41	1447.42	1444.44	1441.47	1438.52	1435.58
490	1432.65	1429.73	1426.82	1423.93	1421.05	1418.18	1415.32	1412.47	1409.63	1406.81
500	1404	1401.19	1398.4	1395.62	1392.85	1390.09	1387.35	1384.61	1381.88	1379.17

TABLE 6-2
3/4-WAVELENGTH VERTICAL
540 - 1605 KILOHERTZ
USE: AM BROADCASTING

Measurements, in feet, advance in 1 kHz increments from left to right across columns.

kHz	feet	feet	feet	feet	feet	feet	feet	feet	feet	feet
540	1300	1288.07	1276.36	1264.86	1253.57	1242.47	1231.57	1220.86	1210.34	1200
590	1189.83	1179.83	1170	1160.33	1150.81	1141.46	1132.25	1123.19	1114.28	1105.51
640	1096.87	1088.37	1080	1071.75	1063.63	1055.63	1047.76	1040	1032.35	1024.81
690	1017.39	1010.07	1002.85	995.74	988.73	981.81	975	968.27	961.64	955.1
740	948.64	942.28	936	929.8	923.68	917.64	911.68	905.8	900	894.26
790	888.6	883.01	877.5	872.04	866.66	861.34	856.09	850.9	845.78	840.71
840	835.71	830.76	825.88	821.05	816.27	811.56	806.89	802.28	797.72	793.22
890	788.76	784.35	780	775.69	771.42	767.21	763.04	758.91	754.83	750.8
940	746.8	742.85	738.94	735.07	731.25	727.46	723.71	720	716.32	712.69
990	709.09	705.52	702	698.5	695.04	691.62	688.23	684.87	681.55	678.26
1040	675	671.77	668.57	665.4	662.26	659.15	656.07	653.02	650	647
1090	644.03	641.09	638.18	635.29	632.43	629.59	626.78	624	621.23	618.5
1140	615.78	613.1	610.43	607.79	605.17	602.57	600	597.44	594.91	592.4
1190	589.91	587.44	585	582.57	580.16	577.77	575.4	573.06	570.73	568.42
1240	566.12	563.85	561.59	559.36	557.14	554.94	552.75	550.58	548.43	546.3
1290	544.18	542.08	540	537.93	535.87	533.84	531.81	529.81	527.81	525.84
1340	523.88	521.93	520	518.08	516.17	514.28	512.4	510.54	508.69	506.85
1390	505.03	503.22	501.42	499.64	497.87	496.11	494.36	492.63	490.9	489.19
1440	487.5	485.81	484.13	482.47	480.82	479.18	477.55	475.93	474.32	472.72
1490	471.14	469.56	468	466.44	464.9	463.36	461.84	460.32	458.82	457.32
1540	455.84	454.36	452.9	451.44	450	448.56	447.13	445.71	444.3	442.9
1590	441.5	440.12	438.75	437.38	436.02	434.67	433.33	432	430.67	429.35

TABLE 6-3
3/4-WAVELENGTH VERTICAL
1800 - 2000 KILOHERTZ
USE: AMATEUR

Measurements, in feet, advance in 1 kHz increments from left to right across columns.

kHz	feet	feet	feet	feet	feet	feet	feet	feet	feet	feet
1800	390	389.78	389.56	389.35	389.13	388.91	388.7	388.48	388.27	388.05
1810	387.84	387.63	387.41	387.2	386.99	386.77	386.56	386.35	386.13	385.92
1820	385.71	385.5	385.29	385.07	384.86	384.65	384.44	384.23	384.02	383.81
1830	383.6	383.39	383.18	382.97	382.76	382.56	382.35	382.14	381.93	381.72
1840	381.52	381.31	381.1	380.9	380.69	380.48	380.28	380.07	379.87	379.66
1850	379.45	379.25	379.04	378.84	378.64	378.43	378.23	378.02	377.82	377.62
1860	377.41	377.21	377.01	376.81	376.6	376.4	376.2	376	375.8	375.6
1870	375.4	375.2	375	374.79	374.59	374.4	374.2	374	373.8	373.6
1880	373.4	373.2	373	372.8	372.61	372.41	372.21	372.01	371.82	371.62
1890	371.42	371.23	371.03	370.83	370.64	370.44	370.25	370.05	369.86	369.66
1900	369.47	369.27	369.08	368.89	368.69	368.5	368.31	368.11	367.92	367.73
1910	367.53	367.34	367.15	366.96	366.77	366.57	366.38	366.19	366	365.81
1920	365.62	365.43	365.24	365.05	364.86	364.67	364.48	364.29	364.1	363.91
1930	363.73	363.54	363.35	363.16	362.97	362.79	362.6	362.41	362.22	362.04
1940	361.85	361.66	361.48	361.29	361.11	360.92	360.73	360.55	360.36	360.18
1950	360	359.81	359.63	359.44	359.26	359.07	358.89	358.71	358.52	358.34
1960	358.16	357.98	357.79	357.61	357.43	357.25	357.07	356.88	356.7	356.52
1970	356.34	356.16	355.98	355.8	355.62	355.44	355.26	355.08	354.9	354.72
1980	354.54	354.36	354.18	354	353.83	353.65	353.47	353.29	353.11	352.94
1990	352.76	352.58	352.4	352.23	352.05	351.87	351.7	351.52	351.35	351.17
2000	351	350.82	350.64	350.47	350.29	350.12	349.95	349.77	349.6	349.42

TABLE 6-4
3/4-WAVELENGTH VERTICAL
3500 - 4000 KILOHERTZ
USE: AMATEUR

Measurements, in feet, advance in 1 kHz increments from left to right across columns.

kHz	feet	feet	feet	feet	feet	feet	feet	feet	feet	feet
3500	200.57	200.45	200.34	200.22	200.11	200	199.88	199.77	199.65	199.54
3520	199.43	199.31	199.2	199.09	198.97	198.86	198.75	198.64	198.52	198.41
3540	198.3	198.19	198.08	197.96	197.85	197.74	197.63	197.52	197.41	197.3
3560	197.19	197.08	196.96	196.85	196.74	196.63	196.52	196.41	196.3	196.19
3580	196.08	195.97	195.87	195.76	195.65	195.54	195.43	195.32	195.21	195.1
3600	195	194.89	194.78	194.67	194.56	194.45	194.35	194.24	194.13	194.02
3620	193.92	193.81	193.7	193.6	193.49	193.38	193.28	193.17	193.06	192.96
3640	192.85	192.75	192.64	192.53	192.43	192.32	192.22	192.11	192.01	191.9
3660	191.8	191.69	191.59	191.48	191.38	191.28	191.17	191.07	190.96	190.86
3680	190.76	190.65	190.55	190.45	190.34	190.24	190.14	190.03	189.93	189.83
3700	189.72	189.62	189.52	189.42	189.32	189.21	189.11	189.01	188.91	188.81
3720	188.7	188.6	188.5	188.4	188.3	188.2	188.1	188	187.9	187.8
3740	187.7	187.6	187.5	187.39	187.29	187.2	187.1	187	186.9	186.8
3760	186.7	186.6	186.5	186.4	186.3	186.2	186.1	186	185.91	185.81
3780	185.71	185.61	185.51	185.41	185.32	185.22	185.12	185.02	184.93	184.83
3800	184.73	184.63	184.54	184.44	184.34	184.25	184.15	184.05	183.96	183.86
3820	183.76	183.67	183.57	183.48	183.38	183.28	183.19	183.09	183	182.9
3840	182.81	182.71	182.62	182.52	182.43	182.33	182.24	182.14	182.05	181.95
3860	181.86	181.77	181.67	181.58	181.48	181.39	181.3	181.2	181.11	181.02
3880	180.92	180.83	180.74	180.64	180.55	180.46	180.36	180.27	180.18	180.09
3900	180	179.9	179.81	179.72	179.63	179.53	179.44	179.35	179.26	179.17
3920	179.08	178.99	178.89	178.8	178.71	178.62	178.53	178.44	178.35	178.26
3940	178.17	178.08	177.99	177.9	177.81	177.72	177.63	177.54	177.45	177.36
3960	177.27	177.18	177.09	177	176.91	176.82	176.73	176.64	176.55	176.47
3980	176.38	176.29	176.2	176.11	176.02	175.93	175.85	175.76	175.67	175.58
4000	175.5	175.41	175.32	175.23	175.14	175.06	174.97	174.88	174.8	174.71

TABLE 6-5
3/4-WAVELENGTH VERTICAL
5500 - 6000 KILOHERTZ
USE: INT'L. BROADCAST

Measurements, in feet, advance in 1 kHz increments from left to right across columns.

kHz	feet	feet	feet	feet	feet	feet	feet	feet	feet	feet
5500	127.63	127.58	127.54	127.49	127.45	127.4	127.35	127.31	127.26	127.22
5520	127.17	127.12	127.08	127.03	126.98	126.94	126.89	126.85	126.8	126.76
5540	126.71	126.66	126.62	126.57	126.53	126.48	126.44	126.39	126.34	126.3
5560	126.25	126.21	126.16	126.12	126.07	126.03	125.98	125.94	125.89	125.85
5580	125.8	125.76	125.71	125.67	125.62	125.58	125.53	125.49	125.44	125.4
5600	125.35	125.31	125.26	125.22	125.17	125.13	125.08	125.04	125	124.95
5620	124.91	124.86	124.82	124.77	124.73	124.68	124.64	124.6	124.55	124.51
5640	124.46	124.42	124.37	124.33	124.29	124.24	124.2	124.15	124.11	124.07
5660	124.02	123.98	123.94	123.89	123.85	123.8	123.76	123.72	123.67	123.63
5680	123.59	123.54	123.5	123.46	123.41	123.37	123.33	123.28	123.24	123.2
5700	123.15	123.11	123.07	123.02	122.98	122.94	122.89	122.85	122.81	122.77
5720	122.72	122.68	122.64	122.59	122.55	122.51	122.47	122.42	122.38	122.34
5740	122.29	122.25	122.21	122.17	122.12	122.08	122.04	122	121.95	121.91
5760	121.87	121.83	121.79	121.74	121.7	121.66	121.62	121.57	121.53	121.49
5780	121.45	121.41	121.36	121.32	121.28	121.24	121.2	121.15	121.11	121.07
5800	121.03	120.99	120.95	120.9	120.86	120.82	120.78	120.74	120.7	120.66
5820	120.61	120.57	120.53	120.49	120.45	120.41	120.37	120.32	120.28	120.24
5840	120.2	120.16	120.12	120.08	120.04	120	119.95	119.91	119.87	119.83
5860	119.79	119.75	119.71	119.67	119.63	119.59	119.55	119.5	119.46	119.42
5880	119.38	119.34	119.3	119.26	119.22	119.18	119.14	119.1	119.06	119.02
5900	118.98	118.94	118.9	118.86	118.82	118.78	118.74	118.7	118.66	118.62
5920	118.58	118.54	118.5	118.46	118.42	118.38	118.34	118.3	118.26	118.22
5940	118.18	118.14	118.1	118.06	118.02	117.98	117.94	117.9	117.86	117.82
5960	117.78	117.74	117.7	117.66	117.62	117.58	117.54	117.5	117.46	117.43
5980	117.39	117.35	117.31	117.27	117.23	117.19	117.15	117.11	117.07	117.03
6000	117	116.96	116.92	116.88	116.84	116.8	116.76	116.72	116.68	116.65

TABLE 6-6
3/4-WAVELENGTH VERTICAL
6000 - 6500 KILOHERTZ
USE: INT'L. BROADCAST

Measurements, in feet, advance in 1 kHz increments from left to right across columns.

kHz	feet	feet	feet	feet	feet	feet	feet	feet	feet	feet
6000	117	116.96	116.92	116.88	116.84	116.8	116.76	116.72	116.68	116.65
6020	116.61	116.57	116.53	116.49	116.45	116.41	116.37	116.34	116.3	116.26
6040	116.22	116.18	116.14	116.1	116.07	116.03	115.99	115.95	115.91	115.87
6060	115.84	115.8	115.76	115.72	115.68	115.65	115.61	115.57	115.53	115.49
6080	115.46	115.42	115.38	115.34	115.3	115.27	115.23	115.19	115.15	115.11
6100	115.08	115.04	115	114.96	114.93	114.89	114.85	114.81	114.78	114.74
6120	114.7	114.66	114.63	114.59	114.55	114.51	114.48	114.44	114.4	114.36
6140	114.33	114.29	114.25	114.22	114.18	114.14	114.1	114.07	114.03	113.99
6160	113.96	113.92	113.88	113.85	113.81	113.77	113.73	113.7	113.66	113.62
6180	113.59	113.55	113.51	113.48	113.44	113.4	113.37	113.33	113.29	113.26
6200	113.22	113.18	113.15	113.11	113.07	113.04	113	112.97	112.93	112.89
6220	112.86	112.82	112.78	112.75	112.71	112.68	112.64	112.6	112.57	112.53
6240	112.5	112.46	112.42	112.39	112.35	112.32	112.28	112.24	112.21	112.17
6260	112.14	112.1	112.06	112.03	111.99	111.96	111.92	111.89	111.85	111.81
6280	111.78	111.74	111.71	111.67	111.64	111.6	111.57	111.53	111.49	111.46
6300	111.42	111.39	111.35	111.32	111.28	111.25	111.21	111.18	111.14	111.11
6320	111.07	111.04	111	110.97	110.93	110.9	110.86	110.83	110.79	110.76
6340	110.72	110.69	110.65	110.62	110.58	110.55	110.51	110.48	110.44	110.41
6360	110.37	110.34	110.3	110.27	110.23	110.2	110.16	110.13	110.1	110.06
6380	110.03	109.99	109.96	109.92	109.89	109.85	109.82	109.79	109.75	109.72
6400	109.68	109.65	109.61	109.58	109.55	109.51	109.48	109.44	109.41	109.37
6420	109.34	109.31	109.27	109.24	109.2	109.17	109.14	109.1	109.07	109.04
6440	109	108.97	108.93	108.9	108.87	108.83	108.8	108.76	108.73	108.7
6460	108.66	108.63	108.6	108.56	108.53	108.5	108.46	108.43	108.4	108.36
6480	108.33	108.29	108.26	108.23	108.19	108.16	108.13	108.09	108.06	108.03
6500	108	107.96	107.93	107.9	107.86	107.83	107.8	107.76	107.73	107.7

TABLE 6-7
3/4-WAVELENGTH VERTICAL
7000 - 7300 KILOHERTZ
USE: AMATEUR

Measurements, in feet, advance in 1 kHz increments from left to right across columns.

kHz	feet	feet	feet	feet	feet	feet	feet	feet	feet	feet
7000	100.28	100.25	100.22	100.19	100.17	100.14	100.11	100.08	100.05	100.02
7020	100	99.97	99.94	99.91	99.88	99.85	99.82	99.8	99.77	99.74
7040	99.71	99.68	99.65	99.63	99.6	99.57	99.54	99.51	99.48	99.46
7060	99.43	99.4	99.37	99.34	99.32	99.29	99.26	99.23	99.2	99.18
7080	99.15	99.12	99.09	99.06	99.04	99.01	98.98	98.95	98.92	98.9
7100	98.87	98.84	98.81	98.78	98.76	98.73	98.7	98.67	98.65	98.62
7120	98.59	98.56	98.54	98.51	98.48	98.45	98.42	98.4	98.37	98.34
7140	98.31	98.29	98.26	98.23	98.2	98.18	98.15	98.12	98.09	98.07
7160	98.04	98.01	97.98	97.96	97.93	97.9	97.88	97.85	97.82	97.79
7180	97.77	97.74	97.71	97.68	97.66	97.63	97.6	97.58	97.55	97.52
7200	97.5	97.47	97.44	97.41	97.39	97.36	97.33	97.31	97.28	97.25
7220	97.22	97.2	97.17	97.14	97.12	97.09	97.06	97.04	97.01	96.98
7240	96.96	96.93	96.9	96.88	96.85	96.82	96.8	96.77	96.74	96.72
7260	96.69	96.66	96.64	96.61	96.58	96.56	96.53	96.5	96.48	96.45
7280	96.42	96.4	96.37	96.34	96.32	96.29	96.26	96.24	96.21	96.19
7300	96.16	96.13	96.11	96.08	96.05	96.03	96	95.98	95.95	95.92

TABLE 6-8
3/4-WAVELENGTH VERTICAL
9500 - 10000 KILOHERTZ
USE: INT'L. BROADCAST

Measurements, in feet, advance in 1 kHz increments from left to right across columns.

kHz	feet	feet	feet	feet	feet	feet	feet	feet	feet	feet
9500	73.89	73.87	73.86	73.84	73.83	73.81	73.8	73.78	73.77	73.75
9520	73.73	73.72	73.7	73.69	73.67	73.66	73.64	73.63	73.61	73.6
9540	73.58	73.56	73.55	73.53	73.52	73.5	73.49	73.47	73.46	73.44
9560	73.43	73.41	73.4	73.38	73.36	73.35	73.33	73.32	73.3	73.29
9580	73.27	73.26	73.24	73.23	73.21	73.2	73.18	73.17	73.15	73.14
9600	73.12	73.1	73.09	73.07	73.06	73.04	73.03	73.01	73	72.98
9620	72.97	72.95	72.94	72.92	72.91	72.89	72.88	72.86	72.85	72.83
9640	72.82	72.8	72.79	72.77	72.76	72.74	72.73	72.71	72.7	72.68
9660	72.67	72.65	72.64	72.62	72.61	72.59	72.58	72.56	72.55	72.53
9680	72.52	72.5	72.49	72.47	72.46	72.44	72.43	72.41	72.4	72.38
9700	72.37	72.35	72.34	72.32	72.31	72.29	72.28	72.26	72.25	72.23
9720	72.22	72.2	72.19	72.17	72.16	72.14	72.13	72.11	72.1	72.08
9740	72.07	72.05	72.04	72.02	72.01	72	71.98	71.97	71.95	71.94
9760	71.92	71.91	71.89	71.88	71.86	71.85	71.83	71.82	71.8	71.79
9780	71.77	71.76	71.74	71.73	71.72	71.7	71.69	71.67	71.66	71.64
9800	71.63	71.61	71.6	71.58	71.57	71.55	71.54	71.53	71.51	71.5
9820	71.48	71.47	71.45	71.44	71.42	71.41	71.39	71.38	71.37	71.35
9840	71.34	71.32	71.31	71.29	71.28	71.26	71.25	71.24	71.22	71.21
9860	71.19	71.18	71.16	71.15	71.13	71.12	71.11	71.09	71.08	71.06
9880	71.05	71.03	71.02	71	70.99	70.98	70.96	70.95	70.93	70.92
9900	70.9	70.89	70.88	70.86	70.85	70.83	70.82	70.8	70.79	70.78
9920	70.76	70.75	70.73	70.72	70.7	70.69	70.68	70.66	70.65	70.63
9940	70.62	70.6	70.59	70.58	70.56	70.55	70.53	70.52	70.51	70.49
9960	70.48	70.46	70.45	70.43	70.42	70.41	70.39	70.38	70.36	70.35
9980	70.34	70.32	70.31	70.29	70.28	70.27	70.25	70.24	70.22	70.21
10000	70.19	70.18	70.17	70.15	70.14	70.12	70.11	70.1	70.08	70.07

TABLE 6-9
3/4-WAVELENGTH VERTICAL
11500 - 12000 KILOHERTZ
USE: INT'L. BROADCAST

Measurements, in feet, advance in 1 kHz increments from left to right across columns.

kHz	feet	feet	feet	feet	feet	feet	feet	feet	feet	feet
11500	61.04	61.03	61.02	61.01	61	60.99	60.97	60.96	60.95	60.94
11520	60.93	60.92	60.91	60.9	60.89	60.88	60.87	60.86	60.85	60.84
11540	60.83	60.82	60.81	60.8	60.78	60.77	60.76	60.75	60.74	60.73
11560	60.72	60.71	60.7	60.69	60.68	60.67	60.66	60.65	60.64	60.63
11580	60.62	60.61	60.6	60.59	60.57	60.56	60.55	60.54	60.53	60.52
11600	60.51	60.5	60.49	60.48	60.47	60.46	60.45	60.44	60.43	60.42
11620	60.41	60.4	60.39	60.38	60.37	60.36	60.35	60.34	60.33	60.31
11640	60.3	60.29	60.28	60.27	60.26	60.25	60.24	60.23	60.22	60.21
11660	60.2	60.19	60.18	60.17	60.16	60.15	60.14	60.13	60.12	60.11
11680	60.1	60.09	60.08	60.07	60.06	60.05	60.04	60.03	60.02	60.01
11700	60	59.98	59.97	59.96	59.95	59.94	59.93	59.92	59.91	59.9
11720	59.89	59.88	59.87	59.86	59.85	59.84	59.83	59.82	59.81	59.8
11740	59.79	59.78	59.77	59.76	59.75	59.74	59.73	59.72	59.71	59.7
11760	59.69	59.68	59.67	59.66	59.65	59.64	59.63	59.62	59.61	59.6
11780	59.59	59.58	59.57	59.56	59.55	59.54	59.53	59.52	59.51	59.5
11800	59.49	59.48	59.47	59.46	59.45	59.44	59.43	59.42	59.41	59.4
11820	59.39	59.38	59.37	59.36	59.35	59.34	59.33	59.32	59.31	59.3
11840	59.29	59.28	59.27	59.26	59.25	59.24	59.23	59.22	59.21	59.2
11860	59.19	59.18	59.17	59.16	59.15	59.14	59.13	59.12	59.11	59.1
11880	59.09	59.08	59.07	59.06	59.05	59.04	59.03	59.02	59.01	59
11900	58.99	58.98	58.97	58.96	58.95	58.94	58.93	58.92	58.91	58.9
11920	58.89	58.88	58.87	58.86	58.85	58.84	58.83	58.82	58.81	58.8
11940	58.79	58.78	58.77	58.76	58.75	58.74	58.73	58.72	58.71	58.7
11960	58.69	58.68	58.67	58.66	58.65	58.64	58.63	58.62	58.61	58.6
11980	58.59	58.58	58.57	58.56	58.55	58.54	58.53	58.52	58.51	58.5
12000	58.5	58.49	58.48	58.47	58.46	58.45	58.44	58.43	58.42	58.41

TABLE 6-10
3/4-WAVELENGTH VERTICAL
14000 - 14450 KILOHERTZ
USE: AMATEUR

Measurements, in feet, advance in 1 kHz increments from left to right across columns.

kHz	feet	feet	feet	feet	feet	feet	feet	feet	feet	feet
14000	50.14	50.13	50.12	50.12	50.11	50.1	50.09	50.09	50.08	50.07
14020	50.07	50.06	50.05	50.04	50.04	50.03	50.02	50.02	50.01	50
14040	50	49.99	49.98	49.97	49.97	49.96	49.95	49.95	49.94	49.93
14060	49.92	49.92	49.91	49.9	49.9	49.89	49.88	49.87	49.87	49.86
14080	49.85	49.85	49.84	49.83	49.82	49.82	49.81	49.8	49.8	49.79
14100	49.78	49.78	49.77	49.76	49.75	49.75	49.74	49.73	49.73	49.72
14120	49.71	49.7	49.7	49.69	49.68	49.68	49.67	49.66	49.66	49.65
14140	49.64	49.63	49.63	49.62	49.61	49.61	49.6	49.59	49.59	49.58
14160	49.57	49.56	49.56	49.55	49.54	49.54	49.53	49.52	49.52	49.51
14180	49.5	49.49	49.49	49.48	49.47	49.47	49.46	49.45	49.45	49.44
14200	49.43	49.42	49.42	49.41	49.4	49.4	49.39	49.38	49.38	49.37
14220	49.36	49.36	49.35	49.34	49.33	49.33	49.32	49.31	49.31	49.3
14240	49.29	49.29	49.28	49.27	49.27	49.26	49.25	49.24	49.24	49.23
14260	49.22	49.22	49.21	49.2	49.2	49.19	49.18	49.18	49.17	49.16
14280	49.15	49.15	49.14	49.13	49.13	49.12	49.11	49.11	49.1	49.09
14300	49.09	49.08	49.07	49.07	49.06	49.05	49.04	49.04	49.03	49.02
14320	49.02	49.01	49	49	48.99	48.98	48.98	48.97	48.96	48.96
14340	46.95	48.94	48.94	48.93	48.92	48.91	48.91	48.9	48.89	48.89
14360	48.88	48.87	48.87	48.86	48.85	48.85	48.84	48.83	48.83	48.82
14380	48.81	48.81	48.8	48.79	48.79	48.78	48.77	48.77	48.76	48.75
14400	48.75	48.74	48.73	48.72	48.72	48.71	48.7	48.7	48.69	48.68
14420	48.68	48.67	48.66	48.66	48.65	48.64	48.64	48.63	48.62	48.62
14440	48.61	48.6	48.6	48.59	48.58	48.58	48.57	48.56	48.56	48.55

TABLE 6-11
3/4-WAVELENGTH VERTICAL
15500 - 16000 KILOHERTZ
USE: INT'L. BROADCAST

Measurements, in feet, advance in 1 kHz increments from left to right across columns.

kHz	feet	feet	feet	feet	feet	feet	feet	feet	feet	feet
15500	45.29	45.28	45.27	45.27	45.26	45.26	45.25	45.24	45.24	45.23
15520	45.23	45.22	45.22	45.21	45.2	45.2	45.19	45.19	45.18	45.17
15540	45.17	45.16	45.16	45.15	45.15	45.14	45.13	45.13	45.12	45.12
15560	45.11	45.1	45.1	45.09	45.09	45.08	45.08	45.07	45.06	45.06
15580	45.05	45.05	45.04	45.04	45.03	45.02	45.02	45.01	45.01	45
15600	45	44.99	44.98	44.98	44.97	44.97	44.96	44.95	44.95	44.94
15620	44.94	44.93	44.93	44.92	44.91	44.91	44.9	44.9	44.89	44.89
15640	44.88	44.87	44.87	44.86	44.86	44.85	44.85	44.84	44.83	44.83
15660	44.82	44.82	44.81	44.81	44.8	44.79	44.79	44.78	44.78	44.77
15680	44.77	44.76	44.75	44.75	44.74	44.74	44.73	44.73	44.72	44.71
15700	44.71	44.7	44.7	44.69	44.69	44.68	44.67	44.67	44.66	44.66
15720	44.65	44.65	44.64	44.63	44.63	44.62	44.62	44.61	44.61	44.6
15740	44.59	44.59	44.58	44.58	44.57	44.57	44.56	44.56	44.55	44.54
15760	44.54	44.53	44.53	44.52	44.52	44.51	44.5	44.5	44.49	44.49
15780	44.48	44.48	44.47	44.46	44.46	44.45	44.45	44.44	44.44	44.43
15800	44.43	44.42	44.41	44.41	44.4	44.4	44.39	44.39	44.38	44.37
15820	44.37	44.36	44.36	44.35	44.35	44.34	44.34	44.33	44.32	44.32
15840	44.31	44.31	44.3	44.3	44.29	44.29	44.28	44.27	44.27	44.26
15860	44.26	44.25	44.25	44.24	44.23	44.23	44.22	44.22	44.21	44.21
15880	44.2	44.2	44.19	44.18	44.18	44.17	44.17	44.16	44.16	44.15
15900	44.15	44.14	44.13	44.13	44.12	44.12	44.11	44.11	44.1	44.1
15920	44.09	44.08	44.08	44.07	44.07	44.06	44.06	44.05	44.05	44.04
15940	44.04	44.03	44.02	44.02	44.01	44.01	44	44	43.99	43.99
15960	43.98	43.97	43.97	43.96	43.96	43.95	43.95	43.94	43.94	43.93
15980	43.92	43.92	43.91	43.91	43.9	43.9	43.89	43.89	43.88	43.88
16000	43.87	43.86	43.86	43.85	43.85	43.84	43.84	43.83	43.83	43.82

TABLE 6-12
3/4-WAVELENGTH VERTICAL
16500 - 17000 KILOHERTZ
USE: INT'L. BROADCAST

Measurements, in feet, advance in 1 kHz increments from left to right across columns.

kHz	feet	feet	feet	feet	feet	feet	feet	feet	feet	feet
16500	42.54	42.54	42.53	42.52	42.52	42.51	42.51	42.5	42.5	42.49
16520	42.49	42.48	42.48	42.47	42.47	42.46	42.46	42.45	42.45	42.44
16540	42.44	42.43	42.43	42.42	42.42	42.41	42.41	42.4	42.4	42.39
16560	42.39	42.38	42.38	42.37	42.37	42.36	42.36	42.35	42.35	42.34
16580	42.34	42.33	42.32	42.32	42.31	42.31	42.3	42.3	42.29	42.29
16600	42.28	42.28	42.27	42.27	42.26	42.26	42.25	42.25	42.24	42.24
16620	42.23	42.23	42.22	42.22	42.21	42.21	42.2	42.2	42.19	42.19
16640	42.18	42.18	42.17	42.17	42.16	42.16	42.15	42.15	42.14	42.14
16660	42.13	42.13	42.12	42.12	42.11	42.11	42.1	42.1	42.09	42.09
16680	42.08	42.08	42.07	42.07	42.06	42.06	42.05	42.05	42.04	42.04
16700	42.03	42.03	42.02	42.02	42.01	42.01	42	42	41.99	41.99
16720	41.98	41.98	41.97	41.97	41.96	41.96	41.95	41.95	41.94	41.94
16740	41.93	41.93	41.92	41.92	41.91	41.91	41.9	41.9	41.89	41.89
16760	41.88	41.88	41.87	41.87	41.86	41.86	41.85	41.85	41.84	41.84
16780	41.83	41.83	41.82	41.82	41.81	41.81	41.8	41.8	41.79	41.79
16800	41.78	41.78	41.77	41.77	41.76	41.76	41.75	41.75	41.74	41.74
16820	41.73	41.73	41.72	41.72	41.71	41.71	41.7	41.7	41.69	41.69
16840	41.68	41.68	41.67	41.67	41.66	41.66	41.65	41.65	41.64	41.64
16860	41.63	41.63	41.62	41.62	41.61	41.61	41.6	41.6	41.59	41.59
16880	41.58	41.58	41.57	41.57	41.56	41.56	41.55	41.55	41.54	41.54
16900	41.53	41.53	41.52	41.52	41.51	41.51	41.5	41.5	41.49	41.49
16920	41.48	41.48	41.47	41.47	41.46	41.46	41.45	41.45	41.45	41.44
16940	41.44	41.43	41.43	41.42	41.42	41.41	41.41	41.4	41.4	41.39
16960	41.39	41.38	41.38	41.37	41.37	41.36	41.35	41.35	41.35	41.34
16980	41.34	41.33	41.33	41.32	41.32	41.31	41.31	41.3	41.3	41.29
17000	41.29	41.28	41.28	41.27	41.27	41.26	41.26	41.26	41.25	41.25

TABLE 6-13
3/4-WAVELENGTH VERTICAL
18500 - 19000 KILOHERTZ
USE: INT'L. BROADCAST

Measurements, in feet, advance in 1 kHz increments from left to right across columns.

kHz	feet	feet	feet	feet	feet	feet	feet	feet	feet	feet
18500	37.94	37.94	37.93	37.93	37.92	37.92	37.92	37.91	37.91	37.9
18520	37.9	37.9	37.89	37.89	37.88	37.88	37.88	37.87	37.87	37.86
18540	37.86	37.85	37.85	37.85	37.84	37.84	37.83	37.83	37.83	37.82
18560	37.82	37.81	37.81	37.81	37.8	37.8	37.79	37.79	37.79	37.78
18580	37.78	37.77	37.77	37.77	37.76	37.76	37.75	37.75	37.75	37.74
18600	37.74	37.73	37.73	37.72	37.72	37.72	37.71	37.71	37.7	37.7
18620	37.7	37.69	37.69	37.68	37.68	37.68	37.67	37.67	37.66	37.66
18640	37.66	37.65	37.65	37.64	37.64	37.64	37.63	37.63	37.62	37.62
18660	37.62	37.61	37.61	37.6	37.6	37.6	37.59	37.59	37.58	37.58
18680	37.58	37.57	37.57	37.56	37.56	37.56	37.55	37.55	37.54	37.54
18700	37.54	37.53	37.53	37.52	37.52	37.52	37.51	37.51	37.5	37.5
18720	37.5	37.49	37.49	37.48	37.48	37.47	37.47	37.47	37.46	37.46
18740	37.45	37.45	37.45	37.44	37.44	37.44	37.43	37.43	37.42	37.42
18760	37.42	37.41	37.41	37.4	37.4	37.4	37.39	37.39	37.38	37.38
18780	37.38	37.37	37.37	37.36	37.36	37.36	37.35	37.35	37.34	37.34
18800	37.34	37.33	37.33	37.32	37.32	37.32	37.31	37.31	37.3	37.3
18820	37.3	37.29	37.29	37.28	37.28	37.28	37.27	37.27	37.26	37.26
18840	37.26	37.25	37.25	37.24	37.24	37.24	37.23	37.23	37.22	37.22
18860	37.22	37.21	37.21	37.2	37.2	37.2	37.19	37.19	37.19	37.18
18880	37.18	37.17	37.17	37.17	37.16	37.16	37.15	37.15	37.15	37.14
18900	37.14	37.13	37.13	37.13	37.12	37.12	37.11	37.11	37.11	37.1
18920	37.1	37.09	37.09	37.09	37.08	37.08	37.08	37.07	37.07	37.06
18940	37.06	37.06	37.05	37.05	37.04	37.04	37.04	37.03	37.03	37.02
18960	37.02	37.02	37.01	37.01	37	37	37	36.99	36.99	36.99
18980	36.98	36.98	36.97	36.97	36.97	36.96	36.96	36.95	36.95	36.95
19000	36.94	36.94	36.93	36.93	36.93	36.92	36.92	36.92	36.91	36.91

TABLE 6-14
3/4-WAVELENGTH VERTICAL
21000 - 21450 KILOHERTZ
USE: AMATEUR

Measurements, in feet, advance in 1 kHz increments from left to right across columns.

kHz	feet	feet	feet	feet	feet	feet	feet	feet	feet	feet
21000	33.42	33.42	33.42	33.41	33.41	33.41	33.4	33.4	33.4	33.39
21020	33.39	33.39	33.39	33.38	33.38	33.38	33.37	33.37	33.37	33.36
21040	33.36	33.36	33.35	33.35	33.35	33.34	33.34	33.34	33.33	33.33
21060	33.33	33.33	33.32	33.32	33.32	33.31	33.31	33.31	33.3	33.3
21080	33.3	33.29	33.29	33.29	33.28	33.28	33.28	33.27	33.27	33.27
21100	33.27	33.26	33.26	33.26	33.25	33.25	33.25	33.24	33.24	33.24
21120	33.23	33.23	33.23	33.22	33.22	33.22	33.21	33.21	33.21	33.21
21140	33.2	33.2	33.2	33.19	33.19	33.19	33.18	33.18	33.18	33.17
21160	33.17	33.17	33.16	33.16	33.16	33.16	33.15	33.15	33.15	33.14
21180	33.14	33.14	33.13	33.13	33.13	33.12	33.12	33.12	33.11	33.11
21200	33.11	33.11	33.1	33.1	33.1	33.09	33.09	33.09	33.08	33.08
21220	33.08	33.07	33.07	33.07	33.06	33.06	33.06	33.06	33.05	33.05
21240	33.05	33.04	33.04	33.04	33.03	33.03	33.03	33.02	33.02	33.02
21260	33.01	33.01	33.01	33.01	33	33	33	32.99	32.99	32.99
21280	32.98	32.98	32.98	32.97	32.97	32.97	32.97	32.96	32.96	32.96
21300	32.95	32.95	32.95	32.94	32.94	32.94	32.93	32.93	32.93	32.92
21320	32.92	32.92	32.92	32.91	32.91	32.91	32.9	32.9	32.9	32.89
21340	32.89	32.89	32.88	32.88	32.88	32.88	32.87	32.87	32.87	32.86
21360	32.86	32.86	32.85	32.85	32.85	32.84	32.84	32.84	32.84	32.83
21380	32.83	32.83	32.82	32.82	32.82	32.81	32.81	32.81	32.8	32.8
21400	32.8	32.8	32.79	32.79	32.79	32.78	32.78	32.78	32.77	32.77
21420	32.77	32.77	32.76	32.76	32.76	32.75	32.75	32.75	32.74	32.74
21440	32.74	32.73	32.73	32.73	32.73	32.72	32.72	32.72	32.71	32.71

TABLE 6-15
3/4-WAVELENGTH VERTICAL
25500 - 26000 KILOHERTZ
USE: INT'L. BROADCAST

Measurements, in feet, advance in 1 kHz increments from left to right across columns.

kHz	feet	feet	feet	feet	feet	feet	feet	feet	feet	feet
25500	27.52	27.52	27.52	27.52	27.52	27.51	27.51	27.51	27.51	27.5
25520	27.5	27.5	27.5	27.5	27.49	27.49	27.49	27.49	27.49	27.48
25540	27.48	27.48	27.48	27.47	27.47	27.47	27.47	27.47	27.46	27.46
25560	27.46	27.46	27.46	27.45	27.45	27.45	27.45	27.44	27.44	27.44
25580	27.44	27.44	27.43	27.43	27.43	27.43	27.43	27.42	27.42	27.42
25600	27.42	27.41	27.41	27.41	27.41	27.41	27.4	27.4	27.4	27.4
25620	27.4	27.39	27.39	27.39	27.39	27.38	27.38	27.38	27.38	27.38
25640	27.37	27.37	27.37	27.37	27.37	27.36	27.36	27.36	27.36	27.35
25660	27.35	27.35	27.35	27.35	27.34	27.34	27.34	27.34	27.34	27.33
25680	27.33	27.33	27.33	27.33	27.32	27.32	27.32	27.32	27.31	27.31
25700	27.31	27.31	27.31	27.3	27.3	27.3	27.3	27.3	27.29	27.29
25720	27.29	27.29	27.28	27.28	27.28	27.28	27.28	27.27	27.27	27.27
25740	27.27	27.27	27.26	27.26	27.26	27.26	27.26	27.25	27.25	27.25
25760	27.25	27.24	27.24	27.24	27.24	27.24	27.23	27.23	27.23	27.23
25780	27.23	27.22	27.22	27.22	27.22	27.21	27.21	27.21	27.21	27.21
25800	27.2	27.2	27.2	27.2	27.2	27.19	27.19	27.19	27.19	27.19
25820	27.18	27.18	27.18	27.18	27.17	27.17	27.17	27.17	27.17	27.16
25840	27.16	27.16	27.16	27.16	27.15	27.15	27.15	27.15	27.15	27.14
25860	27.14	27.14	27.14	27.13	27.13	27.13	27.13	27.13	27.12	27.12
25880	27.12	27.12	27.12	27.11	27.11	27.11	27.11	27.11	27.1	27.1
25900	27.1	27.1	27.1	27.09	27.09	27.09	27.09	27.08	27.08	27.08
25920	27.08	27.08	27.07	27.07	27.07	27.07	27.07	27.06	27.06	27.06
25940	27.06	27.06	27.05	27.05	27.05	27.05	27.04	27.04	27.04	27.04
25960	27.04	27.03	27.03	27.03	27.03	27.03	27.02	27.02	27.02	27.02
25980	27.02	27.01	27.01	27.01	27.01	27.01	27	27	27	27
26000	27	26.99	26.99	26.99	26.99	26.98	26.98	26.98	26.98	26.98

TABLE 6-16
3/4-WAVELENGTH VERTICAL
26500 - 27000 KILOHERTZ
USE: INT'L. BROADCAST

Measurements, in feet, advance in 1 kHz increments from left to right across columns.

kHz	feet	feet	feet	feet	feet	feet	feet	feet	feet	feet
26500	26.49	26.48	26.48	26.48	26.48	26.48	26.47	26.47	26.47	26.47
26520	26.47	26.46	26.46	26.46	26.46	26.46	26.45	26.45	26.45	26.45
26540	26.45	26.44	26.44	26.44	26.44	26.44	26.43	26.43	26.43	26.43
26560	26.43	26.42	26.42	26.42	26.42	26.42	26.41	26.41	26.41	26.41
26580	26.41	26.4	26.4	26.4	26.4	26.4	26.39	26.39	26.39	26.39
26600	26.39	26.38	26.38	26.38	26.38	26.38	26.37	26.37	26.37	26.37
26620	26.37	26.36	26.36	26.36	26.36	26.36	26.35	26.35	26.35	26.35
26640	26.35	26.34	26.34	26.34	26.34	26.34	26.33	26.33	26.33	26.33
26660	26.33	26.32	26.32	26.32	26.32	26.32	26.31	26.31	26.31	26.31
26680	26.31	26.3	26.3	26.3	26.3	26.3	26.3	26.29	26.29	26.29
26700	26.29	26.29	26.28	26.28	26.28	26.28	26.28	26.27	26.27	26.27
26720	26.27	26.27	26.26	26.26	26.26	26.26	26.26	26.25	26.25	26.25
26740	26.25	26.25	26.24	26.24	26.24	26.24	26.24	26.23	26.23	26.23
26760	26.23	26.23	26.22	26.22	26.22	26.22	26.22	26.21	26.21	26.21
26780	26.21	26.21	26.2	26.2	26.2	26.2	26.2	26.19	26.19	26.19
26800	26.19	26.19	26.19	26.18	26.18	26.18	26.18	26.18	26.17	26.17
26820	26.17	26.17	26.17	26.16	26.16	26.16	26.16	26.16	26.15	26.15
26840	26.15	26.15	26.15	26.14	26.14	26.14	26.14	26.14	26.13	26.13
26860	26.13	26.13	26.13	26.12	26.12	26.12	26.12	26.12	26.11	26.11
26880	26.11	26.11	26.11	26.11	26.1	26.1	26.1	26.1	26.1	26.09
26900	26.09	26.09	26.09	26.09	26.08	26.08	26.08	26.08	26.08	26.07
26920	26.07	26.07	26.07	26.07	26.06	26.06	26.06	26.06	26.06	26.05
26940	26.05	26.05	26.05	26.05	26.05	26.04	26.04	26.04	26.04	26.04
26960	26.03	26.03	26.03	26.03	26.03	26.02	26.02	26.02	26.02	26.02
26980	26.01	26.01	26.01	26.01	26.01	26	26	26	26	26
27000	26	25.99	25.99	25.99	25.99	25.99	25.98	25.98	25.98	25.98

TABLE 6-17
3/4-WAVELENGTH VERTICAL
28500 - 29000 KILOHERTZ
USE: AMATEUR

Measurements, in feet, advance in 1 kHz increments from left to right across columns.

kHz	feet	feet	feet	feet	feet	feet	feet	feet	feet	feet
28500	24.63	24.62	24.62	24.62	24.62	24.62	24.62	24.61	24.61	24.61
28520	24.61	24.61	24.61	24.6	24.6	24.6	24.6	24.6	24.6	24.59
28540	24.59	24.59	24.59	24.59	24.59	24.58	24.58	24.58	24.58	24.58
28560	24.57	24.57	24.57	24.57	24.57	24.57	24.56	24.56	24.56	24.56
28580	24.56	24.56	24.55	24.55	24.55	24.55	24.55	24.55	24.54	24.54
28600	24.54	24.54	24.54	24.54	24.53	24.53	24.53	24.53	24.53	24.53
28620	24.52	24.52	24.52	24.52	24.52	24.51	24.51	24.51	24.51	24.51
28640	24.51	24.5	24.5	24.5	24.5	24.5	24.5	24.49	24.49	24.49
28660	24.49	24.49	24.49	24.48	24.48	24.48	24.48	24.48	24.48	24.47
28680	24.47	24.47	24.47	24.47	24.47	24.46	24.46	24.46	24.46	24.46
28700	24.45	24.45	24.45	24.45	24.45	24.45	24.44	24.44	24.44	24.44
28720	24.44	24.44	24.43	24.43	24.43	24.43	24.43	24.43	24.42	24.42
28740	24.42	24.42	24.42	24.42	24.41	24.41	24.41	24.41	24.41	24.41
28760	24.4	24.4	24.4	24.4	24.4	24.4	24.39	24.39	24.39	24.39
28780	24.39	24.39	24.38	24.38	24.38	24.38	24.38	24.38	24.37	24.37
28800	24.37	24.37	24.37	24.36	24.36	24.36	24.36	24.36	24.36	24.35
28820	24.35	24.35	24.35	24.35	24.35	24.34	24.34	24.34	24.34	24.34
28840	24.34	24.33	24.33	24.33	24.33	24.33	24.33	24.32	24.32	24.32
28860	24.32	24.32	24.32	24.31	24.31	24.31	24.31	24.31	24.31	24.3
28880	24.3	24.3	24.3	24.3	24.3	24.29	24.29	24.29	24.29	24.29
28900	24.29	24.28	24.28	24.28	24.28	24.28	24.28	24.27	24.27	24.27
28920	24.27	24.27	24.27	24.26	24.26	24.26	24.26	24.26	24.26	24.25
28940	24.25	24.25	24.25	24.25	24.25	24.24	24.24	24.24	24.24	24.24
28960	24.24	24.23	24.23	24.23	24.23	24.23	24.23	24.22	24.22	24.22
28980	24.22	24.22	24.22	24.21	24.21	24.21	24.21	24.21	24.21	24.2
29000	24.2	24.2	24.2	24.2	24.2	24.19	24.19	24.19	24.19	24.19

TABLE 6-18
3/4-WAVELENGTH VERTICAL
29000 - 29500 KILOHERTZ
USE: AMATEUR

Measurements, in feet, advance in 1 kHz increments from left to right across columns.

kHz	feet	feet	feet	feet	feet	feet	feet	feet	feet	feet
29000	24.2	24.2	24.2	24.2	24.2	24.19	24.19	24.19	24.19	24.19
29020	24.19	24.18	24.18	24.18	24.18	24.18	24.18	24.17	24.17	24.17
29040	24.17	24.17	24.17	24.16	24.16	24.16	24.16	24.16	24.16	24.15
29060	24.15	24.15	24.15	24.15	24.15	24.14	24.14	24.14	24.14	24.14
29080	24.14	24.13	24.13	24.13	24.13	24.13	24.13	24.12	24.12	24.12
29100	24.12	24.12	24.12	24.11	24.11	24.11	24.11	24.11	24.11	24.1
29120	24.1	24.1	24.1	24.1	24.1	24.09	24.09	24.09	24.09	24.09
29140	24.09	24.08	24.08	24.08	24.08	24.08	24.08	24.07	24.07	24.07
29160	24.07	24.07	24.07	24.06	24.06	24.06	24.06	24.06	24.06	24.05
29180	24.05	24.05	24.05	24.05	24.05	24.04	24.04	24.04	24.04	24.04
29200	24.04	24.03	24.03	24.03	24.03	24.03	24.03	24.02	24.02	24.02
29220	24.02	24.02	24.02	24.01	24.01	24.01	24.01	24.01	24.01	24
29240	24	24	24	24	24	24	23.99	23.99	23.99	23.99
29260	23.99	23.99	23.98	23.98	23.98	23.98	23.98	23.98	23.97	23.97
29280	23.97	23.97	23.97	23.97	23.96	23.96	23.96	23.96	23.96	23.96
29300	23.95	23.95	23.95	23.95	23.95	23.95	23.94	23.94	23.94	23.94
29320	23.94	23.94	23.93	23.93	23.93	23.93	23.93	23.93	23.92	23.92
29340	23.92	23.92	23.92	23.92	23.91	23.91	23.91	23.91	23.91	23.91
29360	23.91	23.9	23.9	23.9	23.9	23.9	23.9	23.89	23.89	23.89
29380	23.89	23.89	23.89	23.88	23.88	23.88	23.88	23.88	23.88	23.87
29400	23.87	23.87	23.87	23.87	23.87	23.86	23.86	23.86	23.86	23.86
29420	23.86	23.85	23.85	23.85	23.85	23.85	23.85	23.84	23.84	23.84
29440	23.84	23.84	23.84	23.84	23.83	23.83	23.83	23.83	23.83	23.83
29460	23.82	23.82	23.82	23.82	23.82	23.82	23.81	23.81	23.81	23.81
29480	23.81	23.81	23.8	23.8	23.8	23.8	23.8	23.8	23.79	23.79
29500	23.79	23.79	23.79	23.79	23.79	23.78	23.78	23.78	23.78	23.78

TABLE 6-19
3/4-WAVELENGTH VERTICAL
29500 - 30000 KILOHERTZ
USE: AMATEUR

Measurements, in feet, advance in 1 kHz increments from left to right across columns.

kHz	feet	feet	feet	feet	feet	feet	feet	feet	feet	feet
29500	23.79	23.79	23.79	23.79	23.79	23.78	23.78	23.78	23.78	23.78
29520	23.78	23.77	23.77	23.77	23.77	23.77	23.77	23.76	23.76	23.76
29540	23.76	23.76	23.76	23.75	23.75	23.75	23.75	23.75	23.75	23.74
29560	23.74	23.74	23.74	23.74	23.74	23.74	23.73	23.73	23.73	23.73
29580	23.73	23.73	23.72	23.72	23.72	23.72	23.72	23.72	23.71	23.71
29600	23.71	23.71	23.71	23.71	23.7	23.7	23.7	23.7	23.7	23.7
29620	23.7	23.69	23.69	23.69	23.69	23.69	23.69	23.68	23.68	23.68
29640	23.68	23.68	23.68	23.67	23.67	23.67	23.67	23.67	23.67	23.66
29660	23.66	23.66	23.66	23.66	23.66	23.66	23.65	23.65	23.65	23.65
29680	23.65	23.65	23.64	23.64	23.64	23.64	23.64	23.64	23.63	23.63
29700	23.63	23.63	23.63	23.63	23.62	23.62	23.62	23.62	23.62	23.62
29720	23.62	23.61	23.61	23.61	23.61	23.61	23.61	23.6	23.6	23.6
29740	23.6	23.6	23.6	23.59	23.59	23.59	23.59	23.59	23.59	23.59
29760	23.58	23.58	23.58	23.58	23.58	23.58	23.57	23.57	23.57	23.57
29780	23.57	23.57	23.56	23.56	23.56	23.56	23.56	23.56	23.56	23.55
29800	23.55	23.55	23.55	23.55	23.55	23.54	23.54	23.54	23.54	23.54
29820	23.54	23.53	23.53	23.53	23.53	23.53	23.53	23.53	23.52	23.52
29840	23.52	23.52	23.52	23.52	23.51	23.51	23.51	23.51	23.51	23.51
29860	23.5	23.5	23.5	23.5	23.5	23.5	23.5	23.49	23.49	23.49
29880	23.49	23.49	23.49	23.48	23.48	23.48	23.48	23.48	23.48	23.47
29900	23.47	23.47	23.47	23.47	23.47	23.47	23.46	23.46	23.46	23.46
29920	23.46	23.46	23.45	23.45	23.45	23.45	23.45	23.45	23.45	23.44
29940	23.44	23.44	23.44	23.44	23.44	23.43	23.43	23.43	23.43	23.43
29960	23.43	23.42	23.42	23.42	23.42	23.42	23.42	23.42	23.41	23.41
29980	23.41	23.41	23.41	23.41	23.4	23.4	23.4	23.4	23.4	23.4
30000	23.4	23.39	23.39	23.39	23.39	23.39	23.39	23.38	23.38	23.38

Chapter 7
Yagi Antennas

A "gain" antenna is one that seems to give you something for nothing (it doesn't—actually, it merely redirects your efforts). Both CBers and amateurs know that the use of a gain antenna such as a Yagi, quad, or delta loop will increase their signal strength at a distant station. These antennas work this seeming miracle by focusing, or redirecting the energy from the transmitter so that it is radiated in only one direction. The total amount of energy radiated is the same; after all, the antenna is a passive device and supplies no energy of its own.

We usually measure the gain of antennas by comparing the signal strength at a distant location with that of a dipole mounted close to the antenna being tested, or by a theoretical concept called an isotropic source (i.e. a point source). The difference in signal strength received is then measured and described using *decibel notation.* Hence the practice of saying a certain antenna has "so-many dB gain."

But be very careful of manufacturer's claims as to antenna gain! This is an area, especially in the CB market, that has been subject to a lot of creative specification writing. I have seen advertisements for four-element CB Yagi-type beams which claimed over 25-dB gain. The truth probably lies in the 5-10 dB range, if you are lucky.

Also, the difference between using a dipole reference and an isotropic reference is just under 2 dB, with the isotropic producing the higher figures more advantageous to the sales person.

The beam antenna also provides advantages on receive as well. The antenna will tend to receive in only one general direction, and there might be a 20-30 dB difference between the response of the antenna when pointed away and then pointed toward a distant station. Amateur and CB operators have long used the beam antenna to cut out some of the interference from other stations. If you point the beam antenna toward the station that you wish to talk to, then the strength of interfering stations from other directions is greatly reduced. This has especially good application on the crowded amateur HF and CB bands, where a little power will travel a long way.

Perhaps less popular among shortwave listeners (SWL), the beam antenna can also serve them as well. The same problem of overcrowding plaques the high-frequency international broadcast bands as in the CB and ham bands. The serious SWL may want to construct one of the beam antennas discussed in Chapters 7, 8 and 9 so that interference can be cut drastically.

The basic Yagi beam antenna is shown in Fig. 7-1. There will always be a driven element, which will be either a simple center-fed, half-wavelength dipole, or a folded dipole (see Chapter 1). The other elements are called parasitic elements because they are not directly energized by the transmitter. There are two forms of parasitic elements: directors and reflectors.

The director is shorter than the driven element and is positioned toward the direction of radiation—in other words, toward the other station. The director is usually four to six percent shorter than the half-wavelength driven element.

The reflector is located on the other side of the driven element, the side away from the direction of radiation. The physical length of the reflector is four to six percent longer than the driven element.

In theory, you can have as many reflectors and directors as you want. Most three-element Yagis, the most popular type of beam antenna, use a driven element and one each of the reflector and director. If you want to make a four-element beam antenna, for higher gain than a three-element model, then add either an additional director or an additional reflector.

The directors and reflectors are mounted from 0.1 to 0.25 wavelengths from the driven element, or from each other if more

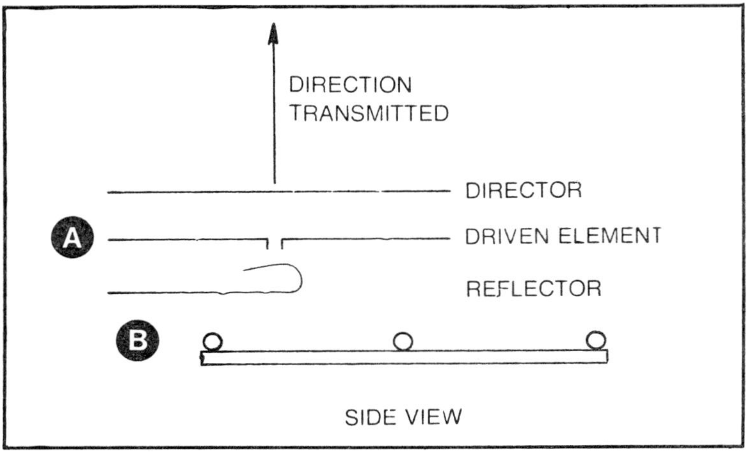

Fig. 7-1. Yagi beam antenna with top view at (A) and side view at (B).

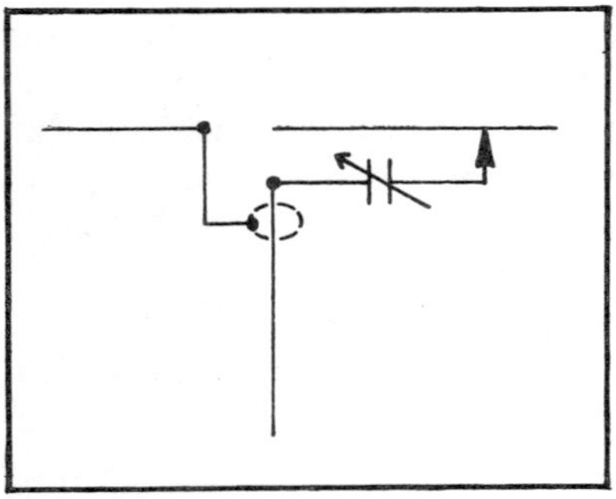

Fig. 7-2 Gamma-match feeding of the Yagi

than one of either type is used. There is an optimal spacing for these elements, however. Many people are of the opinion that the use of more elements actually gives greater gain, as long as the spacing between them is within these ranges. But that is not always true. There is an optimum number of elements that can be accommodated on any given size boom (the boom supports the elements, and its total length is called the "boom length," naturally enough). If additional elements are added, then the gain may actually go down instead of increase as expected. The study of optimizing the design of Yagi antennas could fill several books, and many articles have been written on that subject in the amateur periodicals and book literature. The reader who wants to get more involved is directed to that literature. The tables in this chapter are based on a simple three-element Yagi.

The dimensions for the elements are obtained from the following expressions:

Driven element:	L = 468/F
Director:	L = 450/F
Reflector:	L = 492/F

where L is the length in feet, and F is the frequency in megahertz.

The nominal impedance of a dipole antenna is 70–75 ohms. When parasitic elements are added, however, the impedance will drop. The traditional solution to feeding such an antenna is to use a gamma matching element (Fig. 7-2) to connect the center conductor of the coaxial cable to a higher impedance point along the length of the driven element. More recently, however, some amateurs have replaced the gamma section or its balanced cousin, the T-section, with a toroidal rf transformer. Some of these are available in wideband models (3 – 30 mHz), and will accommodate 6 to 10 impedance levels down to 5 ohms with a 52-ohm feedline.

TABLE 7-1
THREE-ELEMENT YAGI BEAM
3500 - 4000 KILOHERTZ
USE: AMATEUR

Measurements for element lengths and wavelength spacing are in feet.

kHz	FORMAT: DIR./DR.EL./REFL./	SPACING 0.1,	0.125,	0.15,	0.20	0.25
3500	128.5 / 133.7 / 140.5	28.1	35.125	42.15	56.2	70.25
3520	127.8 / 132.9 / 139.7	27.94	34.925	41.91	55.88	69.85
3540	127.1 / 132.2 / 138.9	27.78	34.725	41.67	55.56	69.45
3560	126.4 / 131.4 / 138.2	27.64	34.55	41.46	55.28	69.1
3580	125.6 / 130.7 / 137.4	27.48	34.35	41.22	54.96	68.7
3600	125 / 130 / 136.6	27.32	34.15	40.98	54.64	68.3
3620	124.3 / 129.2 / 135.9	27.18	33.975	40.77	54.36	67.95
3640	123.6 / 128.5 / 135.1	27.02	33.775	40.53	54.04	67.55
3660	122.9 / 127.8 / 134.4	26.88	33.6	40.32	53.76	67.2
3680	122.2 / 127.1 / 133.6	26.72	33.4	40.08	53.44	66.8
3700	121.6 / 126.4 / 132.9	26.58	33.225	39.87	53.16	66.45
3720	120.9 / 125.8 / 132.2	26.44	33.05	39.66	52.88	66.1
3740	120.3 / 125.1 / 131.5	26.3	32.875	39.45	52.6	65.75
3760	119.6 / 124.4 / 130.8	26.16	32.7	39.24	52.32	65.4
3780	119 / 123.8 / 130.1	26.02	32.525	39.03	52.04	65.05
3800	118.4 / 123.1 / 129.4	25.88	32.35	38.82	51.76	64.7
3820	117.8 / 122.5 / 128.7	25.74	32.175	38.61	51.48	64.35
3840	117.1 / 121.8 / 128.1	25.62	32.025	38.43	51.24	64.05
3860	116.5 / 121.2 / 127.4	25.48	31.85	38.22	50.96	63.7
3880	115.9 / 120.6 / 126.8	25.36	31.7	38.04	50.72	63.4
3900	115.3 / 120 / 126.1	25.22	31.525	37.83	50.44	63.05
3920	114.7 / 119.3 / 125.5	25.1	31.375	37.65	50.2	62.75
3940	114.2 / 118.7 / 124.8	24.96	31.2	37.44	49.92	62.4
3960	113.6 / 118.1 / 124.2	24.84	31.05	37.26	49.68	62.1
3980	113 / 117.5 / 123.6	24.72	30.9	37.08	49.44	61.8
4000	112.5 / 117 / 123	24.6	30.75	36.9	49.2	61.5

TABLE 7-2
THREE-ELEMENT YAGI BEAM
5500 - 6000 KILOHERTZ
USE: INT'L. BROADCAST

Measurements for element lengths and wavelength spacing are in feet.

kHz	FORMAT: DIR./DR.EL./REFL./	SPACING 0.1,	0.125,	0.15,	0.20	0.25
5500	81.8 / 85 / 89.4	17.88	22.35	26.82	35.76	44.7
5520	81.5 / 84.7 / 89.1	17.82	22.275	26.73	35.64	44.55
5540	81.2 / 84.4 / 88.8	17.76	22.2	26.64	35.52	44.4
5560	80.9 / 84.1 / 88.4	17.68	22.1	26.52	35.36	44.2
5580	80.6 / 83.8 / 88.1	17.62	22.025	26.43	35.24	44.05
5600	80.3 / 83.5 / 87.8	17.56	21.95	26.34	35.12	43.9
5620	80 / 83.2 / 87.5	17.5	21.875	26.25	35	43.75
5640	79.7 / 82.9 / 87.2	17.44	21.8	26.16	34.88	43.6
5660	79.5 / 82.6 / 86.9	17.38	21.725	26.07	34.76	43.45
5680	79.2 / 82.3 / 86.6	17.32	21.65	25.98	34.64	43.3
5700	78.9 / 82.1 / 86.3	17.26	21.575	25.89	34.52	43.15
5720	78.6 / 81.8 / 86	17.2	21.5	25.8	34.4	43
5740	78.3 / 81.5 / 85.7	17.14	21.425	25.71	34.28	42.85
5760	78.1 / 81.2 / 85.4	17.08	21.35	25.62	34.16	42.7
5780	77.8 / 80.9 / 85.1	17.02	21.275	25.53	34.04	42.55
5800	77.5 / 80.6 / 84.8	16.96	21.2	25.44	33.92	42.4
5820	77.3 / 80.4 / 84.5	16.9	21.125	25.35	33.8	42.25
5840	77 / 80.1 / 84.2	16.84	21.05	25.26	33.68	42.1
5860	76.7 / 79.8 / 83.7	16.78	20.975	25.17	33.56	41.95
5880	76.5 / 79.5 / 83.6	16.72	20.9	25.08	33.44	41.8
5900	76.2 / 79.3 / 83.3	16.66	20.825	24.99	33.32	41.65
5920	76 / 79 / 83.1	16.62	20.775	24.93	33.24	41.55
5940	75.7 / 78.7 / 82.8	16.56	20.7	24.84	33.12	41.4
5960	75.5 / 78.5 / 82.5	16.5	20.625	24.75	33	41.25
5980	75.2 / 78.2 / 82.2	16.44	20.55	24.66	32.88	41.1
6000	75 / 78 / 82	16.4	20.5	24.6	32.8	41

TABLE 7-3
THREE-ELEMENT YAGI BEAM
6000 - 6500 KILOHERTZ
USE: INT'L. BROADCAST

Measurements for element lengths and wavelength spacing are in feet.

kHz	FORMAT: DIR./DR.EL./REFL./	SPACING 0.1,	0.125,	0.15,	0.20	0.25
6000	75 / 78 / 82	16.4	20.5	24.6	32.8	41
6020	74.7 / 77.7 / 81.7	16.34	20.425	24.51	32.68	40.85
6040	74.5 / 77.4 / 81.4	16.28	20.35	24.42	32.56	40.7
6060	74.2 / 77.2 / 81.1	16.22	20.275	24.33	32.44	40.55
6080	74 / 76.9 / 80.9	16.18	20.225	24.27	32.36	40.45
6100	73.7 / 76.7 / 80.6	16.12	20.15	24.18	32.24	40.3
6120	73.5 / 76.4 / 80.3	16.06	20.075	24.09	32.12	40.15
6140	73.2 / 76.2 / 80.1	16.02	20.025	24.03	32.04	40.05
6160	73 / 75.9 / 79.8	15.96	19.95	23.94	31.92	39.9
6180	72.8 / 75.7 / 79.6	15.92	19.9	23.88	31.84	39.8
6200	72.5 / 75.4 / 79.3	15.86	19.825	23.79	31.72	39.65
6220	72.3 / 75.2 / 79	15.8	19.75	23.7	31.6	39.5
6240	72.1 / 75 / 78.8	15.76	19.7	23.64	31.52	39.4
6260	71.8 / 74.7 / 78.5	15.7	19.625	23.55	31.4	39.25
6280	71.6 / 74.5 / 78.3	15.66	19.575	23.49	31.32	39.15
6300	71.4 / 74.2 / 78	15.6	19.5	23.4	31.2	39
6320	71.2 / 74 / 77.8	15.56	19.45	23.34	31.12	38.9
6340	70.9 / 73.8 / 77.6	15.52	19.4	23.28	31.04	38.8
6360	70.7 / 73.5 / 77.3	15.46	19.325	23.19	30.92	38.65
6380	70.5 / 73.3 / 77.1	15.42	19.275	23.13	30.84	38.55
6400	70.3 / 73.1 / 76.8	15.36	19.2	23.04	30.72	38.4
6420	70 / 72.8 / 76.6	15.32	19.15	22.98	30.64	38.3
6440	69.8 / 72.6 / 76.3	15.26	19.075	22.89	30.52	38.15
6460	69.6 / 72.4 / 76.1	15.22	19.025	22.83	30.44	38.05
6480	69.4 / 72.2 / 75.9	15.18	18.975	22.77	30.36	37.95
6500	69.2 / 72 / 75.6	15.12	18.9	22.68	30.24	37.8

TABLE 7-4
THREE-ELEMENT YAGI BEAM
7000 - 7300 KILOHERTZ
USE: AMATEUR

Measurements for element lengths and wavelength spacing are in feet.

kHz	FORMAT: DIR./DR.EL./REFL./	SPACING 0.1,	0.125,	0.15,	0.20	0.25
7000	64.2 / 66.8 / 70.2	14.04	17.55	21.06	28.08	35.1
7020	64.1 / 66.6 / 70	14	17.5	21	28	35
7040	63.9 / 66.4 / 69.8	13.96	17.45	20.94	27.92	34.9
7060	63.7 / 66.2 / 69.6	13.92	17.4	20.88	27.84	34.8
7080	63.5 / 66.1 / 69.4	13.88	17.35	20.82	27.76	34.7
7100	63.3 / 65.9 / 69.2	13.84	17.3	20.76	27.68	34.6
7120	63.2 / 65.7 / 69.1	13.82	17.275	20.73	27.64	34.55
7140	63 / 65.5 / 68.9	13.78	17.225	20.67	27.56	34.45
7160	62.8 / 65.3 / 68.7	13.74	17.175	20.61	27.48	34.35
7180	62.6 / 65.1 / 68.5	13.7	17.125	20.55	27.4	34.25
7200	62.5 / 65 / 68.3	13.66	17.075	20.49	27.32	34.15
7220	62.3 / 64.8 / 68.1	13.62	17.025	20.43	27.24	34.05
7240	62.1 / 64.6 / 67.9	13.58	16.975	20.37	27.16	33.95
7260	61.9 / 64.4 / 67.7	13.54	16.925	20.31	27.08	33.85
7280	61.8 / 64.2 / 67.5	13.5	16.875	20.25	27	33.75
7300	61.6 / 64.1 / 67.3	13.46	16.825	20.19	26.92	33.65

TABLE 7-5
THREE-ELEMENT YAGI BEAM
9500 - 10000 KILOHERTZ
USE: INT'L. BROADCAST

Measurements for element lengths and wavelength spacing are in feet.

kHz	FORMAT: DIR./DR.EL./REFL./	SPACING 0.1,	0.125,	0.15,	0.20	0.25
9500	47.3 / 49.2 / 51.7	10.34	12.925	15.51	20.68	25.85
9520	47.2 / 49.1 / 51.6	10.32	12.9	15.48	20.64	25.8
9540	47.1 / 49 / 51.5	10.3	12.875	15.45	20.6	25.75
9560	47 / 48.9 / 51.4	10.28	12.85	15.42	20.56	25.7
9580	46.9 / 48.8 / 51.3	10.26	12.825	15.39	20.52	25.65
9600	46.8 / 48.7 / 51.2	10.24	12.8	15.36	20.48	25.6
9620	46.7 / 48.6 / 51.1	10.22	12.775	15.33	20.44	25.55
9640	46.6 / 48.5 / 51	10.2	12.75	15.3	20.4	25.5
9660	46.5 / 48.4 / 50.9	10.18	12.725	15.27	20.36	25.45
9680	46.4 / 48.3 / 50.8	10.16	12.7	15.24	20.32	25.4
9700	46.3 / 48.2 / 50.7	10.14	12.675	15.21	20.28	25.35
9720	46.2 / 48.1 / 50.6	10.12	12.65	15.18	20.24	25.3
9740	46.2 / 48 / 50.5	10.1	12.625	15.15	20.2	25.25
9760	46.1 / 47.9 / 50.4	10.08	12.6	15.12	20.16	25.2
9780	46 / 47.8 / 50.3	10.06	12.575	15.09	20.12	25.15
9800	45.9 / 47.7 / 50.2	10.04	12.55	15.06	20.08	25.1
9820	45.8 / 47.6 / 50.1	10.02	12.525	15.03	20.04	25.05
9840	45.7 / 47.5 / 50	10	12.5	15	20	25
9860	45.6 / 47.4 / 49.8	9.96	12.45	14.94	19.92	24.9
9880	45.5 / 47.3 / 49.7	9.94	12.425	14.91	19.88	24.85
9900	45.4 / 47.2 / 49.6	9.92	12.4	14.88	19.84	24.8
9920	45.3 / 47.1 / 49.5	9.9	12.375	14.85	19.8	24.75
9940	45.2 / 47 / 49.4	9.88	12.35	14.82	19.76	24.7
9960	45.1 / 46.9 / 49.3	9.86	12.325	14.79	19.72	24.65
9980	45 / 46.8 / 49.2	9.84	12.3	14.76	19.68	24.6
10000	45 / 46.8 / 49.2	9.84	12.3	14.76	19.68	24.6

TABLE 7-6
THREE-ELEMENT YAGI BEAM
11500 - 12000 KILOHERTZ
USE: INT´L. BROADCAST

Measurements for element lengths and wavelength spacing are in feet.

kHz	FORMAT: DIR./DR.EL./REFL./	SPACING 0.1,	0.125,	0.15,	0.20	0.25
11500	39.1 / 40.6 / 42.7	8.54	10.675	12.81	17.08	21.35
11520	39 / 40.6 / 42.7	8.54	10.675	12.81	17.08	21.35
11540	38.9 / 40.5 / 42.6	8.52	10.65	12.78	17.04	21.3
11560	38.9 / 40.4 / 42.5	8.5	10.625	12.75	17	21.25
11580	38.8 / 40.4 / 42.4	8.48	10.6	12.72	16.96	21.2
11600	38.7 / 40.3 / 42.4	8.48	10.6	12.72	16.96	21.2
11620	38.7 / 40.2 / 42.3	8.46	10.575	12.69	16.92	21.15
11640	38.6 / 40.2 / 42.2	8.44	10.55	12.66	16.88	21.1
11660	38.5 / 40.1 / 42.1	8.42	10.525	12.63	16.84	21.05
11680	38.5 / 40 / 42.1	8.42	10.525	12.63	16.84	21.05
11700	38.4 / 40 / 42	8.4	10.5	12.6	16.8	21
11720	38.3 / 39.9 / 41.9	8.38	10.475	12.57	16.76	20.95
11740	38.3 / 39.8 / 41.9	8.38	10.475	12.57	16.76	20.95
11760	38.2 / 39.7 / 41.8	8.36	10.45	12.54	16.72	20.9
11780	38.2 / 39.7 / 41.7	8.34	10.425	12.51	16.68	20.85
11800	38.1 / 39.6 / 41.6	8.32	10.4	12.48	16.64	20.8
11820	38 / 39.5 / 41.6	8.32	10.4	12.48	16.64	20.8
11840	38 / 39.5 / 41.5	8.3	10.375	12.45	16.6	20.75
11860	37.9 / 39.4 / 41.4	8.28	10.35	12.42	16.56	20.7
11880	37.8 / 39.3 / 41.4	8.28	10.35	12.42	16.56	20.7
11900	37.8 / 39.3 / 41.3	8.26	10.325	12.39	16.52	20.65
11920	37.7 / 39.2 / 41.2	8.24	10.3	12.36	16.48	20.6
11940	37.6 / 39.1 / 41.2	8.24	10.3	12.36	16.48	20.6
11960	37.6 / 39.1 / 41.1	8.22	10.275	12.33	16.44	20.55
11980	37.5 / 39 / 41	8.2	10.25	12.3	16.4	20.5
12000	37.5 / 39 / 41	8.2	10.25	12.3	16.4	20.5

TABLE 7-7
THREE-ELEMENT YAGI BEAM
14000 - 14450 KILOHERTZ
USE: AMATEUR

Measurements for element lengths and wavelength spacing are in feet.

kHz	FORMAT: DIR./DR.EL./REFL./	SPACING 0.1,	0.125,	0.15,	0.20	0.25
14000	32.1 / 33.4 / 35.1	7.02	8.775	10.53	14.04	17.55
14020	32 / 33.3 / 35	7	8.75	10.5	14	17.5
14040	32 / 33.3 / 35	7	8.75	10.5	14	17.5
14060	32 / 33.2 / 34.9	6.98	8.725	10.47	13.96	17.45
14080	31.9 / 33.2 / 34.9	6.98	8.725	10.47	13.96	17.45
14100	31.9 / 33.1 / 34.8	6.96	8.7	10.44	13.92	17.4
14120	31.8 / 33.1 / 34.8	6.96	8.7	10.44	13.92	17.4
14140	31.8 / 33 / 34.7	6.94	8.675	10.41	13.88	17.35
14160	31.7 / 33 / 34.7	6.94	8.675	10.41	13.88	17.35
14180	31.7 / 33 / 34.6	6.92	8.65	10.38	13.84	17.3
14200	31.6 / 32.9 / 34.6	6.92	8.65	10.38	13.84	17.3
14220	31.6 / 32.9 / 34.5	6.9	8.625	10.35	13.8	17.25
14240	31.6 / 32.8 / 34.5	6.9	8.625	10.35	13.8	17.25
14260	31.5 / 32.8 / 34.5	6.9	8.625	10.35	13.8	17.25
14280	31.5 / 32.7 / 34.4	6.88	8.6	10.32	13.76	17.2
14300	31.4 / 32.7 / 34.4	6.88	8.6	10.32	13.76	17.2
14320	31.4 / 32.6 / 34.3	6.86	8.575	10.29	13.72	17.15
14340	31.3 / 32.6 / 34.3	6.86	8.575	10.29	13.72	17.15
14360	31.3 / 32.5 / 34.2	6.84	8.55	10.26	13.68	17.1
14380	31.2 / 32.5 / 34.2	6.84	8.55	10.26	13.68	17.1
14400	31.2 / 32.5 / 34.1	6.82	8.525	10.23	13.64	17.05
14420	31.2 / 32.4 / 34.1	6.82	8.525	10.23	13.64	17.05
14440	31.1 / 32.4 / 34	6.8	8.5	10.2	13.6	17

TABLE 7-8
THREE-ELEMENT YAGI BEAM
15500 - 16000 KILOHERTZ
USE: INT'L. BROADCAST

Measurements for element lengths and wavelength spacing are in feet.

kHz	FORMAT: DIR./DR.EL./REFL./	SPACING 0.1,	0.125,	0.15,	0.20	0.25
15500	29 / 30.1 / 31.7	6.34	7.925	9.51	12.68	15.85
15520	28.9 / 30.1 / 31.7	6.34	7.925	9.51	12.68	15.85
15540	28.9 / 30.1 / 31.6	6.32	7.9	9.48	12.64	15.8
15560	28.9 / 30 / 31.6	6.32	7.9	9.48	12.64	15.8
15580	28.8 / 30 / 31.5	6.3	7.875	9.45	12.6	15.75
15600	28.8 / 30 / 31.5	6.3	7.875	9.45	12.6	15.75
15620	28.8 / 29.9 / 31.4	6.28	7.85	9.42	12.56	15.7
15640	28.7 / 29.9 / 31.4	6.28	7.85	9.42	12.56	15.7
15660	28.7 / 29.8 / 31.4	6.28	7.85	9.42	12.56	15.7
15680	28.6 / 29.8 / 31.3	6.26	7.825	9.39	12.52	15.65
15700	28.6 / 29.8 / 31.3	6.26	7.825	9.39	12.52	15.65
15720	28.6 / 29.7 / 31.2	6.24	7.8	9.36	12.48	15.6
15740	28.5 / 29.7 / 31.2	6.24	7.8	9.36	12.48	15.6
15760	28.5 / 29.6 / 31.2	6.24	7.8	9.36	12.48	15.6
15780	28.5 / 29.6 / 31.1	6.22	7.775	9.33	12.44	15.55
15800	28.4 / 29.6 / 31.1	6.22	7.775	9.33	12.44	15.55
15820	28.4 / 29.5 / 31	6.2	7.75	9.3	12.4	15.5
15840	28.4 / 29.5 / 31	6.2	7.75	9.3	12.4	15.5
15860	28.3 / 29.5 / 31	6.2	7.75	9.3	12.4	15.5
15880	28.3 / 29.4 / 30.9	6.18	7.725	9.27	12.36	15.45
15900	28.3 / 29.4 / 30.9	6.18	7.725	9.27	12.36	15.45
15920	28.2 / 29.3 / 30.9	6.18	7.725	9.27	12.36	15.45
15940	28.2 / 29.3 / 30.8	6.16	7.7	9.24	12.32	15.4
15960	28.1 / 29.3 / 30.8	6.16	7.7	9.24	12.32	15.4
15980	28.1 / 29.2 / 30.7	6.14	7.675	9.21	12.28	15.35
16000	28.1 / 29.2 / 30.7	6.14	7.675	9.21	12.28	15.35

TABLE 7-9
THREE-ELEMENT YAGI BEAM
16500 - 17000 KILOHERTZ
USE: INT'L. BROADCAST

Measurements for element lengths and wavelength spacing are in feet.

kHz	FORMAT: DIR./DR.EL./REFL./	SPACING 0.1,	0.125,	0.15,	0.20	0.25
16500	27.2 / 28.3 / 29.8	5.96	7.45	8.94	11.92	14.9
16520	27.2 / 28.3 / 29.7	5.94	7.425	8.91	11.88	14.85
16540	27.2 / 28.2 / 29.7	5.94	7.425	8.91	11.88	14.85
16560	27.1 / 28.2 / 29.7	5.94	7.425	8.91	11.88	14.85
16580	27.1 / 28.2 / 29.6	5.92	7.4	8.88	11.84	14.8
16600	27.1 / 28.1 / 29.6	5.92	7.4	8.88	11.84	14.8
16620	27 / 28.1 / 29.6	5.92	7.4	8.88	11.84	14.8
16640	27 / 28.1 / 29.5	5.9	7.375	8.85	11.8	14.75
16660	27 / 28 / 29.5	5.9	7.375	8.85	11.8	14.75
16680	26.9 / 28 / 29.4	5.88	7.35	8.82	11.76	14.7
16700	26.9 / 28 / 29.4	5.88	7.35	8.82	11.76	14.7
16720	26.9 / 27.9 / 29.4	5.88	7.35	8.82	11.76	14.7
16740	26.8 / 27.9 / 29.3	5.86	7.325	8.79	11.72	14.65
16760	26.8 / 27.9 / 29.3	5.86	7.325	8.79	11.72	14.65
16780	26.8 / 27.8 / 29.3	5.86	7.325	8.79	11.72	14.65
16800	26.7 / 27.8 / 29.2	5.84	7.3	8.76	11.68	14.6
16820	26.7 / 27.8 / 29.2	5.84	7.3	8.76	11.68	14.6
16840	26.7 / 27.7 / 29.2	5.84	7.3	8.76	11.68	14.6
16860	26.6 / 27.7 / 29.1	5.82	7.275	8.73	11.64	14.55
16880	26.6 / 27.7 / 29.1	5.82	7.275	8.73	11.64	14.55
16900	26.6 / 27.6 / 29.1	5.82	7.275	8.73	11.64	14.55
16920	26.5 / 27.6 / 29	5.8	7.25	8.7	11.6	14.5
16940	26.5 / 27.6 / 29	5.8	7.25	8.7	11.6	14.5
16960	26.5 / 27.5 / 29	5.8	7.25	8.7	11.6	14.5
16980	26.5 / 27.5 / 28.9	5.78	7.225	8.67	11.56	14.45
17000	26.4 / 27.5 / 28.9	5.78	7.225	8.67	11.56	14.45

TABLE 7-10
THREE-ELEMENT YAGI BEAM
18500 - 19000 KILOHERTZ
USE: INT'L. BROADCAST

Measurements for element lengths and wavelength spacing are in feet.

kHz	FORMAT: DIR./DR.EL./REFL./	SPACING 0.1,	0.125,	0.15,	0.20	0.25
18500	24.3 / 25.2 / 26.5	5.3	6.625	7.95	10.6	13.25
18520	24.2 / 25.2 / 26.5	5.3	6.625	7.95	10.6	13.25
18540	24.2 / 25.2 / 26.5	5.3	6.625	7.95	10.6	13.25
18560	24.2 / 25.2 / 26.5	5.3	6.625	7.95	10.6	13.25
18580	24.2 / 25.1 / 26.4	5.28	6.6	7.92	10.56	13.2
18600	24.1 / 25.1 / 26.4	5.28	6.6	7.92	10.56	13.2
18620	24.1 / 25.1 / 26.4	5.28	6.6	7.92	10.56	13.2
18640	24.1 / 25.1 / 26.3	5.26	6.575	7.89	10.52	13.15
18660	24.1 / 25 / 26.3	5.26	6.575	7.89	10.52	13.15
18680	24 / 25 / 26.3	5.26	6.575	7.89	10.52	13.15
18700	24 / 25 / 26.3	5.26	6.575	7.89	10.52	13.15
18720	24 / 25 / 26.2	5.24	6.55	7.86	10.48	13.1
18740	24 / 24.9 / 26.2	5.24	6.55	7.86	10.48	13.1
18760	23.9 / 24.9 / 26.2	5.24	6.55	7.86	10.48	13.1
18780	23.9 / 24.9 / 26.1	5.22	6.525	7.83	10.44	13.05
18800	23.9 / 24.8 / 26.1	5.22	6.525	7.83	10.44	13.05
18820	23.9 / 24.8 / 26.1	5.22	6.525	7.83	10.44	13.05
18840	23.8 / 24.8 / 26.1	5.22	6.525	7.83	10.44	13.05
18860	23.8 / 24.8 / 26	5.2	6.5	7.8	10.4	13
18880	23.8 / 24.7 / 26	5.2	6.5	7.8	10.4	13
18900	23.8 / 24.7 / 26	5.2	6.5	7.8	10.4	13
18920	23.7 / 24.7 / 26	5.2	6.5	7.8	10.4	13
18940	23.7 / 24.7 / 25.9	5.18	6.475	7.77	10.36	12.95
18960	23.7 / 24.6 / 25.9	5.18	6.475	7.77	10.36	12.95
18980	23.7 / 24.6 / 25.9	5.18	6.475	7.77	10.36	12.95
19000	23.6 / 24.6 / 25.8	5.16	6.45	7.74	10.32	12.9

TABLE 7-11
THREE-ELEMENT YAGI BEAM
21000 - 21450 KILOHERTZ
'ISE: AMATEUR

Measurements for element lengths and wavelength spacing are in feet.

kHz	FORMAT: DIR./DR.EL./REFL./	SPACING 0.1,	0.125,	0.15,	0.20	0.25
21000	21.4 / 22.2 / 23.4	4.68	5.85	7.02	9.36	11.7
21020	21.4 / 22.2 / 23.4	4.68	5.85	7.02	9.36	11.7
21040	21.3 / 22.2 / 23.3	4.66	5.825	6.99	9.32	11.65
21060	21.3 / 22.2 / 23.3	4.66	5.825	6.99	9.32	11.65
21080	21.3 / 22.2 / 23.3	4.66	5.825	6.99	9.32	11.65
21100	21.3 / 22.1 / 23.3	4.66	5.825	6.99	9.32	11.65
21120	21.3 / 22.1 / 23.2	4.64	5.8	6.96	9.28	11.6
21140	21.2 / 22.1 / 23.2	4.64	5.8	6.96	9.28	11.6
21160	21.2 / 22.1 / 23.2	4.64	5.8	6.96	9.28	11.6
21180	21.2 / 22 / 23.2	4.64	5.8	6.96	9.28	11.6
21200	21.2 / 22 / 23.2	4.64	5.8	6.96	9.28	11.6
21220	21.2 / 22 / 23.1	4.62	5.775	6.93	9.24	11.55
21240	21.1 / 22 / 23.1	4.62	5.775	6.93	9.24	11.55
21260	21.1 / 22 / 23.1	4.62	5.775	6.93	9.24	11.55
21280	21.1 / 21.9 / 23.1	4.62	5.775	6.93	9.24	11.55
21300	21.1 / 21.9 / 23	4.6	5.75	6.9	9.2	11.5
21320	21.1 / 21.9 / 23	4.6	5.75	6.9	9.2	11.5
21340	21 / 21.9 / 23	4.6	5.75	6.9	9.2	11.5
21360	21 / 21.9 / 23	4.6	5.75	6.9	9.2	11.5
21380	21 / 21.8 / 23	4.6	5.75	6.9	9.2	11.5
21400	21 / 21.8 / 22.9	4.58	5.725	6.87	9.16	11.45
21420	21 / 21.8 / 22.9	4.58	5.725	6.87	9.16	11.45
21440	20.9 / 21.8 / 22.9	4.58	5.725	6.87	9.16	11.45

TABLE 7-12
THREE-ELEMENT YAGI BEAM
25500 - 26000 KILOHERTZ
USE: INT´L. BROADCAST

Measurements for element lengths and wavelength spacing are in feet.

kHz	FORMAT: DIR./DR.EL./REFL./	SPACING 0.1,	0.125,	0.15,	0.20	0.25
25500	17.6 / 18.3 / 19.2	3.84	4.8	5.76	7.68	9.6
25520	17.6 / 18.3 / 19.2	3.84	4.8	5.76	7.68	9.6
25540	17.6 / 18.3 / 19.2	3.84	4.8	5.76	7.68	9.6
25560	17.6 / 18.3 / 19.2	3.84	4.8	5.76	7.68	9.6
25580	17.5 / 18.2 / 19.2	3.84	4.8	5.76	7.68	9.6
25600	17.5 / 18.2 / 19.2	3.84	4.8	5.76	7.68	9.6
25620	17.5 / 18.2 / 19.2	3.84	4.8	5.76	7.68	9.6
25640	17.5 / 18.2 / 19.1	3.82	4.775	5.73	7.64	9.55
25660	17.5 / 18.2 / 19.1	3.82	4.775	5.73	7.64	9.55
25680	17.5 / 18.2 / 19.1	3.82	4.775	5.73	7.64	9.55
25700	17.5 / 18.2 / 19.1	3.82	4.775	5.73	7.64	9.55
25720	17.4 / 18.1 / 19.1	3.82	4.775	5.73	7.64	9.55
25740	17.4 / 18.1 / 19.1	3.82	4.775	5.73	7.64	9.55
25760	17.4 / 18.1 / 19	3.8	4.75	5.7	7.6	9.5
25780	17.4 / 18.1 / 19	3.8	4.75	5.7	7.6	9.5
25800	17.4 / 18.1 / 19	3.8	4.75	5.7	7.6	9.5
25820	17.4 / 18.1 / 19	3.8	4.75	5.7	7.6	9.5
25840	17.4 / 18.1 / 19	3.8	4.75	5.7	7.6	9.5
25860	17.4 / 18 / 19	3.8	4.75	5.7	7.6	9.5
25880	17.3 / 18 / 19	3.8	4.75	5.7	7.6	9.5
25900	17.3 / 18 / 18.9	3.78	4.725	5.67	7.56	9.45
25920	17.3 / 18 / 18.9	3.78	4.725	5.67	7.56	9.45
25940	17.3 / 18 / 18.9	3.78	4.725	5.67	7.56	9.45
25960	17.3 / 18 / 18.9	3.78	4.725	5.67	7.56	9.45
25980	17.3 / 18 / 18.9	3.78	4.725	5.67	7.56	9.45
26000	17.3 / 18 / 18.9	3.78	4.725	5.67	7.56	9.45

TABLE 7-13
THREE-ELEMENT YAGI BEAM
26500 - 27000 KILOHERTZ
USE: INT'L. BROADCAST

Measurements for element lengths and wavelength spacing are in feet.

kHz	FORMAT: DIR./DR.EL./REFL./	SPACING 0.1,	0.125,	0.15,	0.20	0.25
26500	16.9 / 17.6 / 18.5	3.7	4.625	5.55	7.4	9.25
26520	16.9 / 17.6 / 18.5	3.7	4.625	5.55	7.4	9.25
26540	16.9 / 17.6 / 18.5	3.7	4.625	5.55	7.4	9.25
26560	16.9 / 17.6 / 18.5	3.7	4.625	5.55	7.4	9.25
26580	16.9 / 17.6 / 18.5	3.7	4.625	5.55	7.4	9.25
26600	16.9 / 17.5 / 18.4	3.68	4.6	5.52	7.36	9.2
26620	16.9 / 17.5 / 18.4	3.68	4.6	5.52	7.36	9.2
26640	16.8 / 17.5 / 18.4	3.68	4.6	5.52	7.36	9.2
26660	16.8 / 17.5 / 18.4	3.68	4.6	5.52	7.36	9.2
26680	16.8 / 17.5 / 18.4	3.68	4.6	5.52	7.36	9.2
26700	16.8 / 17.5 / 18.4	3.68	4.6	5.52	7.36	9.2
26720	16.8 / 17.5 / 18.4	3.68	4.6	5.52	7.36	9.2
26740	16.8 / 17.5 / 18.3	3.66	4.575	5.49	7.32	9.15
26760	16.8 / 17.4 / 18.3	3.66	4.575	5.49	7.32	9.15
26780	16.8 / 17.4 / 18.3	3.66	4.575	5.49	7.32	9.15
26800	16.7 / 17.4 / 18.3	3.66	4.575	5.49	7.32	9.15
26820	16.7 / 17.4 / 18.3	3.66	4.575	5.49	7.32	9.15
26840	16.7 / 17.4 / 18.3	3.66	4.575	5.49	7.32	9.15
26860	16.7 / 17.4 / 18.3	3.66	4.575	5.49	7.32	9.15
26880	16.7 / 17.4 / 18.3	3.66	4.575	5.49	7.32	9.15
26900	16.7 / 17.3 / 18.2	3.64	4.55	5.46	7.28	9.1
26920	16.7 / 17.3 / 18.2	3.64	4.55	5.46	7.28	9.1
26940	16.7 / 17.3 / 18.2	3.64	4.55	5.46	7.28	9.1
26960	16.6 / 17.3 / 18.2	3.64	4.55	5.46	7.28	9.1
26980	16.6 / 17.3 / 18.2	3.64	4.55	5.46	7.28	9.1
27000	16.6 / 17.3 / 18.2	3.64	4.55	5.46	7.28	9.1

TABLE 7-14
THREE-ELEMENT YAGI BEAM
28500 - 29000 KILOHERTZ
USE: AMATEUR

Measurements for element lengths and wavelength spacing are in feet.

kHz	FORMAT: DIR./DR.EL./REFL./	SPACING 0.1,	0.125,	0.15,	0.20	0.25
28500	15.7 / 16.4 / 17.2	3.44	4.3	5.16	6.88	8.6
28520	15.7 / 16.4 / 17.2	3.44	4.3	5.16	6.88	8.6
28540	15.7 / 16.3 / 17.2	3.44	4.3	5.16	6.88	8.6
28560	15.7 / 16.3 / 17.2	3.44	4.3	5.16	6.88	8.6
28580	15.7 / 16.3 / 17.2	3.44	4.3	5.16	6.88	8.6
28600	15.7 / 16.3 / 17.2	3.44	4.3	5.16	6.88	8.6
28620	15.7 / 16.3 / 17.1	3.42	4.275	5.13	6.84	8.55
28640	15.7 / 16.3 / 17.1	3.42	4.275	5.13	6.84	8.55
28660	15.7 / 16.3 / 17.1	3.42	4.275	5.13	6.84	8.55
28680	15.6 / 16.3 / 17.1	3.42	4.275	5.13	6.84	8.55
28700	15.6 / 16.3 / 17.1	3.42	4.275	5.13	6.84	8.55
28720	15.6 / 16.2 / 17.1	3.42	4.275	5.13	6.84	8.55
28740	15.6 / 16.2 / 17.1	3.42	4.275	5.13	6.84	8.55
28760	15.6 / 16.2 / 17.1	3.42	4.275	5.13	6.84	8.55
28780	15.6 / 16.2 / 17	3.4	4.25	5.1	6.8	8.5
28800	15.6 / 16.2 / 17	3.4	4.25	5.1	6.8	8.5
28820	15.6 / 16.2 / 17	3.4	4.25	5.1	6.8	8.5
28840	15.6 / 16.2 / 17	3.4	4.25	5.1	6.8	8.5
28860	15.5 / 16.2 / 17	3.4	4.25	5.1	6.8	8.5
28880	15.5 / 16.2 / 17	3.4	4.25	5.1	6.8	8.5
28900	15.5 / 16.1 / 17	3.4	4.25	5.1	6.8	8.5
28920	15.5 / 16.1 / 17	3.4	4.25	5.1	6.8	8.5
28940	15.5 / 16.1 / 17	3.4	4.25	5.1	6.8	8.5
28960	15.5 / 16.1 / 16.9	3.38	4.225	5.07	6.76	8.45
28980	15.5 / 16.1 / 16.9	3.38	4.225	5.07	6.76	8.45
29000	15.5 / 16.1 / 16.9	3.38	4.225	5.07	6.76	8.45

TABLE 7-15
29000 - 29500 KILOHERTZ
USE: AMATEUR

Measurements for element lengths and wavelength spacing are in feet

kHz	FORMAT: DIR./DR.EL./REFL.	0.1,	0.125,	0.15,	0.2,	0.25
29000	15.51 / 16.13 / 16.96	3.39	4.24	5.08	6.78	8.48
29020	15.5 / 16.12 / 16.95	3.39	4.23	5.08	6.78	8.47
29040	15.49 / 16.11 / 16.94	3.38	4.23	5.08	6.77	8.47
29060	15.48 / 16.1 / 16.93	3.38	4.23	5.07	6.77	8.46
29080	15.47 / 16.09 / 16.91	3.38	4.22	5.07	6.76	8.45
29100	15.46 / 16.08 / 16.9	3.38	4.22	5.07	6.76	8.45
29120	15.45 / 16.07 / 16.89	3.37	4.22	5.06	6.75	8.44
29140	15.44 / 16.06 / 16.88	3.37	4.22	5.06	6.75	8.44
29160	15.43 / 16.04 / 16.87	3.37	4.21	5.06	6.74	8.43
29180	15.42 / 16.03 / 16.86	3.37	4.21	5.05	6.74	8.43
29200	15.41 / 16.02 / 16.84	3.36	4.21	5.05	6.73	8.42
29220	15.4 / 16.01 / 16.83	3.36	4.2	5.05	6.73	8.41
29240	15.38 / 16 / 16.82	3.36	4.2	5.04	6.73	8.41
29260	15.37 / 15.99 / 16.81	3.36	4.2	5.04	6.72	8.4
29280	15.36 / 15.98 / 16.8	3.36	4.2	5.04	6.72	8.4
29300	15.35 / 15.97 / 16.79	3.35	4.19	5.03	6.71	8.39
29320	15.34 / 15.96 / 16.78	3.35	4.19	5.03	6.71	8.39
29340	15.33 / 15.95 / 16.76	3.35	4.19	5.03	6.7	8.38
29360	15.32 / 15.94 / 16.75	3.35	4.18	5.02	6.7	8.37
29380	15.31 / 15.92 / 16.74	3.34	4.18	5.02	6.69	8.37
29400	15.3 / 15.91 / 16.73	3.34	4.18	5.02	6.69	8.36
29420	15.29 / 15.9 / 16.72	3.34	4.18	5.01	6.68	8.36
29440	15.28 / 15.89 / 16.71	3.34	4.17	5.01	6.68	8.35
29460	15.27 / 15.88 / 16.7	3.34	4.17	5.01	6.68	8.35
29480	15.26 / 15.87 / 16.68	3.33	4.17	5	6.67	8.34
29500	15.25 / 15.86 / 16.67	3.33	4.16	5	6.67	8.33

TABLE 7-16
THREE-ELEMENT YAGI BEAM
29500 - 30000 KILOHERTZ
USE: AMATEUR

Measurements for element lengths and wavelength spacing are in feet.

kHz	FORMAT: DIR./DR.EL./REFL./	SPACING 0.1,	0.125,	0.15,	0.20	0.25
29500	15.2 / 15.8 / 16.6	3.32	4.15	4.98	6.64	8.3
29520	15.2 / 15.8 / 16.6	3.32	4.15	4.98	6.64	8.3
29540	15.2 / 15.8 / 16.6	3.32	4.15	4.98	6.64	8.3
29560	15.2 / 15.8 / 16.6	3.32	4.15	4.98	6.64	8.3
29580	15.2 / 15.8 / 16.6	3.32	4.15	4.98	6.64	8.3
29600	15.2 / 15.8 / 16.6	3.32	4.15	4.98	6.64	8.3
29620	15.1 / 15.8 / 16.6	3.32	4.15	4.98	6.64	8.3
29640	15.1 / 15.7 / 16.5	3.3	4.125	4.95	6.6	8.25
29660	15.1 / 15.7 / 16.5	3.3	4.125	4.95	6.6	8.25
29680	15.1 / 15.7 / 16.5	3.3	4.125	4.95	6.6	8.25
29700	15.1 / 15.7 / 16.5	3.3	4.125	4.95	6.6	8.25
29720	15.1 / 15.7 / 16.5	3.3	4.125	4.95	6.6	8.25
29740	15.1 / 15.7 / 16.5	3.3	4.125	4.95	6.6	8.25
29760	15.1 / 15.7 / 16.5	3.3	4.125	4.95	6.6	8.25
29780	15.1 / 15.7 / 16.5	3.3	4.125	4.95	6.6	8.25
29800	15.1 / 15.7 / 16.5	3.3	4.125	4.95	6.6	8.25
29820	15 / 15.6 / 16.4	3.28	4.1	4.92	6.56	8.2
29840	15 / 15.6 / 16.4	3.28	4.1	4.92	6.56	8.2
29860	15 / 15.6 / 16.4	3.28	4.1	4.92	6.56	8.2
29880	15 / 15.6 / 16.4	3.28	4.1	4.92	6.56	8.2
29900	15 / 15.6 / 16.4	3.28	4.1	4.92	6.56	8.2
29920	15 / 15.6 / 16.4	3.28	4.1	4.92	6.56	8.2
29940	15 / 15.6 / 16.4	3.28	4.1	4.92	6.56	8.2
29960	15 / 15.6 / 16.4	3.28	4.1	4.92	6.56	8.2
29980	15 / 15.6 / 16.4	3.28	4.1	4.92	6.56	8.2
30000	15 / 15.6 / 16.4	3.28	4.1	4.92	6.56	8.2

Chapter 8
Quads

The quad antenna is another form of beam, and is said by its supporters to out perform a Yagi when both antennas are mounted close to the earth's surface. This may well be reason enough to select the quad over other types.

Figure 8-1 shows the basic form of quad antenna. Although a three-element version is shown here, most quad antennas are two-element versions. I suspect that this may be due to the easier mechanical assembly of two-element models. Of course, manufacturers, who have solved these problems, offer multi-element quads. On the Citizens Band, where size is not as much of a problem, there are many four- and six-element quads used.

Although not specifically a "quad," it is possible to use a single loop antenna, or a driven element only, and obtain better results than with a simple dipole. Although many do not believe the gain claims made for this type of antenna, it has certain advantages. One is that it is compact, so it can be rotated to orient with the station you want to work, or receive. The same loop design that gives it compactness also provides a superior freedom from noise generated by nearby electrical devices. Of course, the full size beam would do better, but this is a good alternative for those who cannot afford a full size Yagi or quad.

Interestingly enough, the polarity of the signal in a quad depends upon the *feedpoint*! The polarity of any radio signal is the direction, vertical or horizontal, or something in between, of the *electrical* part of the electromagnetic field radiated by the antenna. All antennas radiate both an electrical field and a magnetic field. The magnetic field will be at right angles to the electrical field. A dipole, mounted horizontal to the earth, will radiate a horizontally polarized signal, while a vertical antenna radiates a vertically polarized signal. If the quad driven element is fed along the bottom side of the square, then the field will be horizontally polarized. But if the driven element is fed at the middle of the vertical edge, then a vertically polarized signal is produced.

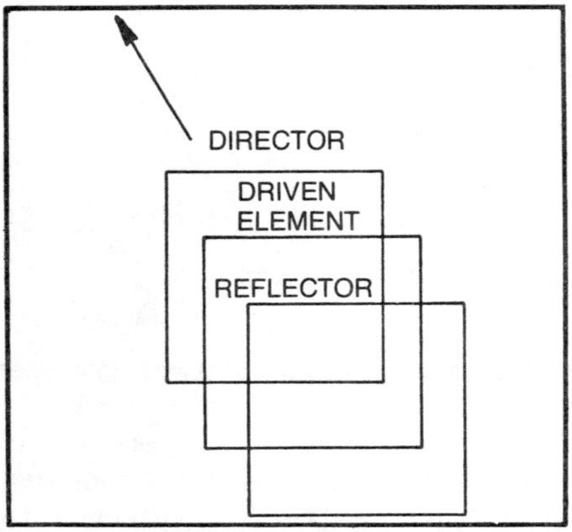

Fig. 8-1. Quad beam antenna.

The loop is in the form of a square, so all sides have equal length. The size of each leg is one-quarter wavelength. The overall length of the wire needed to make the square is one full wavelength, and this is found from

Driven element:	L = 984/F
Reflector:	L = 1030/F
Director:	L = 935/F

In most cases, the loops of a multi-element quad are spaced ⅛-wavelength apart (0.125 wavelength). The spacing, then is found from:

Spacing for 0.125L = 123/F

where L is the length in feet, and F is the frequency in megahertz.

The early quads used by amateurs were constructed of 2×4 and 1×2 lumber to form the support, with bamboo *spreaders* to hold the wire. That was in the day when you could walk into any carpet store and walk out with as many bamboo stalks as you needed, free of charge. But today, most radio amateurs use fiberglass spreaders manufactured by a number of companies. The latest amateur radio magazines usually carry several advertisments for quad antenna parts. Also, many different CB quads are available. The SWL, however, will probably have to use amateur supplies to make a quad because the frequencies of interest to him are more nearly the same as the amateur bands than the CB.

Recently, some amateurs have been using combination quad and Yagi designs in which there is a square loop—as in the quad—which is used as the driven element, and a reflector or director as in the Yagi antenna.

Wire lengths for the quad are given in the following tables.

TABLE 8-1
THREE-ELEMENT QUAD & DELTA-LOOP ANTENNAS
3500 - 4000 KILOHERTZ
USE: AMATEUR

Measurements for element lengths and wavelength spacing are in feet.

kHz	FORMAT: DIR./DR.EL./REFL./	SPACING 0.1,	0.125,	0.15,	0.20	0.25
3500	267.1 / 281.1 / 294.2	28.11	35.1375	42.165	56.22	70.275
3520	265.6 / 279.5 / 292.6	27.95	34.9375	41.925	55.9	69.875
3540	264.1 / 277.9 / 290.9	27.79	34.7375	41.685	55.58	69.475
3560	262.6 / 276.4 / 289.3	27.64	34.55	41.46	55.28	69.1
3580	261.1 / 274.8 / 287.7	27.48	34.35	41.22	54.96	68.7
3600	259.7 / 273.3 / 286.1	27.33	34.1625	40.995	54.66	68.325
3620	258.2 / 271.8 / 284.5	27.18	33.975	40.77	54.36	67.95
3640	256.8 / 270.3 / 282.9	27.03	33.7875	40.545	54.06	67.575
3660	255.4 / 268.8 / 281.4	26.88	33.6	40.32	53.76	67.2
3680	254 / 267.3 / 279.8	26.73	33.4125	40.095	53.46	66.825
3700	252.7 / 265.9 / 278.3	26.59	33.2375	39.885	53.18	66.475
3720	251.3 / 264.5 / 276.8	26.45	33.0625	39.675	52.9	66.125
3740	250 / 263.1 / 275.4	26.31	32.8875	39.465	52.62	65.775
3760	248.6 / 261.7 / 273.9	26.17	32.7125	39.255	52.34	65.425
3780	247.3 / 260.3 / 272.4	26.03	32.5375	39.045	52.06	65.075
3800	246 / 258.9 / 271	25.89	32.3625	38.835	51.78	64.725
3820	244.7 / 257.5 / 269.6	25.75	32.1875	38.625	51.5	64.375
3840	243.4 / 256.2 / 268.2	25.62	32.025	38.43	51.24	64.05
3860	242.2 / 254.9 / 266.8	25.49	31.8625	38.235	50.98	63.725
3880	240.9 / 253.6 / 265.4	25.36	31.7	38.04	50.72	63.4
3900	239.7 / 252.3 / 264.1	25.23	31.5375	37.845	50.46	63.075
3920	238.5 / 251 / 262.7	25.1	31.375	37.65	50.2	62.75
3940	237.3 / 249.7 / 261.4	24.97	31.2125	37.455	49.94	62.425
3960	236.1 / 248.4 / 260.1	24.84	31.05	37.26	49.68	62.1
3980	234.9 / 247.2 / 258.7	24.72	30.9	37.08	49.44	61.8
4000	233.7 / 246 / 257.5	24.6	30.75	36.9	49.2	61.5

TABLE 8-2
THREE-ELEMENT QUAD & DELTA-LOOP ANTENNAS
5500 - 6000 KILOHERTZ
USE: INT'L. BROADCAST

Measurements for element lengths and wavelength spacing are in feet.

kHz	FORMAT: DIR./ DR.EL./ REFL./	SPACING 0.1,	0.125,	0.15,	0.20	0.25
5500	170 / 178.9 / 187.2	17.89	22.3625	26.835	35.78	44.725
5520	169.3 / 178.2 / 186.5	17.82	22.275	26.73	35.64	44.55
5540	168.7 / 177.6 / 185.9	17.76	22.2	26.64	35.52	44.4
5560	168.1 / 176.9 / 185.2	17.69	22.1125	26.535	35.38	44.225
5580	167.5 / 176.3 / 184.5	17.63	22.0375	26.445	35.26	44.075
5600	166.9 / 175.7 / 183.9	17.57	21.9625	26.355	35.14	43.925
5620	166.3 / 175 / 183.2	17.5	21.875	26.25	35	43.75
5640	165.7 / 174.4 / 182.6	17.44	21.8	26.16	34.88	43.6
5660	165.1 / 173.8 / 181.9	17.38	21.725	26.07	34.76	43.45
5680	164.6 / 173.2 / 181.3	17.32	21.65	25.98	34.64	43.3
5700	164 / 172.6 / 180.7	17.26	21.575	25.89	34.52	43.15
5720	163.4 / 172 / 180	17.2	21.5	25.8	34.4	43
5740	162.8 / 171.4 / 179.4	17.14	21.425	25.71	34.28	42.85
5760	162.3 / 170.8 / 178.8	17.08	21.35	25.62	34.16	42.7
5780	161.7 / 170.2 / 178.2	17.02	21.275	25.53	34.04	42.55
5800	161.2 / 169.6 / 177.5	16.96	21.2	25.44	33.92	42.4
5820	160.6 / 169 / 176.9	16.9	21.125	25.35	33.8	42.25
5840	160.1 / 168.4 / 176.3	16.84	21.05	25.26	33.68	42.1
5860	159.5 / 167.9 / 175.7	16.79	20.9875	25.185	33.58	41.975
5880	159 / 167.3 / 175.1	16.73	20.9125	25.095	33.46	41.825
5900	158.4 / 166.7 / 174.5	16.67	20.8375	25.005	33.34	41.675
5920	157.9 / 166.2 / 173.9	16.62	20.775	24.93	33.24	41.55
5940	157.4 / 165.6 / 173.4	16.56	20.7	24.84	33.12	41.4
5960	156.8 / 165.1 / 172.8	16.51	20.6375	24.765	33.02	41.275
5980	156.3 / 164.5 / 172.2	16.45	20.5625	24.675	32.9	41.125
6000	155.8 / 164 / 171.6	16.4	20.5	24.6	32.8	41

TABLE 8-3
THREE-ELEMENT QUAD & DELTA-LOOP ANTENNAS
6000 - 6500 KILOHERTZ
USE: INT'L. BROADCAST

Measurements for element lengths and wavelength spacing are in feet.

kHz	FORMAT: DIR./ DR.EL./ REFL./	SPACING 0.1,	0.125,	0.15,	0.20	0.25
6000	155.8 / 164 / 171.6	16.4	20.5	24.6	32.8	41
6020	155.3 / 163.4 / 171	16.34	20.425	24.51	32.68	40.85
6040	154.8 / 162.9 / 170.5	16.29	20.3625	24.435	32.58	40.725
6060	154.2 / 162.3 / 169.9	16.23	20.2875	24.345	32.46	40.575
6080	153.7 / 161.8 / 169.4	16.18	20.225	24.27	32.36	40.45
6100	153.2 / 161.3 / 168.8	16.13	20.1625	24.195	32.26	40.325
6120	152.7 / 160.7 / 168.3	16.07	20.0875	24.105	32.14	40.175
6140	152.2 / 160.2 / 167.7	16.02	20.025	24.03	32.04	40.05
6160	151.7 / 159.7 / 167.2	15.97	19.9625	23.955	31.94	39.925
6180	151.2 / 159.2 / 166.6	15.92	19.9	23.88	31.84	39.8
6200	150.8 / 158.7 / 166.1	15.87	19.8375	23.805	31.74	39.675
6220	150.3 / 158.1 / 165.5	15.81	19.7625	23.715	31.62	39.525
6240	149.8 / 157.6 / 165	15.76	19.7	23.64	31.52	39.4
6260	149.3 / 157.1 / 164.5	15.71	19.6375	23.565	31.42	39.275
6280	148.8 / 156.6 / 164	15.66	19.575	23.49	31.32	39.15
6300	148.4 / 156.1 / 163.4	15.61	19.5125	23.415	31.22	39.025
6320	147.9 / 155.6 / 162.9	15.56	19.45	23.34	31.12	38.9
6340	147.4 / 155.2 / 162.4	15.52	19.4	23.28	31.04	38.8
6360	147 / 154.7 / 161.9	15.47	19.3375	23.205	30.94	38.675
6380	146.5 / 154.2 / 161.4	15.42	19.275	23.13	30.84	38.55
6400	146 / 153.7 / 160.9	15.37	19.2125	23.055	30.74	38.425
6420	145.6 / 153.2 / 160.4	15.32	19.15	22.98	30.64	38.3
6440	145.1 / 152.7 / 159.9	15.27	19.0875	22.905	30.54	38.175
6460	144.7 / 152.3 / 159.4	15.23	19.0375	22.845	30.46	38.075
6480	144.2 / 151.8 / 158.9	15.18	18.975	22.77	30.36	37.95
6500	143.8 / 151.3 / 158.4	15.13	18.9125	22.695	30.26	37.825

TABLE 8-4
THREE-ELEMENT QUAD & DELTA-LOOP ANTENNAS
7000 - 7300 KILOHERTZ
USE: AMATEUR

Measurements for element lengths and wavelength spacing are in feet.

kHz	FORMAT: DIR./ DR.EL./ REFL./	SPACING 0.1,	0.125,	0.15,	0.20	0.25
7000	133.5 / 140.5 / 147.1	14.05	17.5625	21.075	28.1	35.125
7020	133.1 / 140.1 / 146.7	14.01	17.5125	21.015	28.02	35.025
7040	132.8 / 139.7 / 146.3	13.97	17.4625	20.955	27.94	34.925
7060	132.4 / 139.3 / 145.8	13.93	17.4125	20.895	27.86	34.825
7080	132 / 138.9 / 145.4	13.89	17.3625	20.835	27.78	34.725
7100	131.6 / 138.5 / 145	13.85	17.3125	20.775	27.7	34.625
7120	131.3 / 138.2 / 144.6	13.82	17.275	20.73	27.64	34.55
7140	130.9 / 137.8 / 144.2	13.78	17.225	20.67	27.56	34.45
7160	130.5 / 137.4 / 143.8	13.74	17.175	20.61	27.48	34.35
7180	130.2 / 137 / 143.4	13.7	17.125	20.55	27.4	34.25
7200	129.8 / 136.6 / 143	13.66	17.075	20.49	27.32	34.15
7220	129.5 / 136.2 / 142.6	13.62	17.025	20.43	27.24	34.05
7240	129.1 / 135.9 / 142.2	13.59	16.9875	20.385	27.18	33.975
7260	128.7 / 135.5 / 141.8	13.55	16.9375	20.325	27.1	33.875
7280	128.4 / 135.1 / 141.4	13.51	16.8875	20.265	27.02	33.775
7300	128 / 134.7 / 141	13.47	16.8375	20.205	26.94	33.675

TABLE 8-5
THREE-ELEMENT QUAD & DELTA-LOOP ANTENNAS
9500 - 10000 KILOHERTZ
USE: INT´L. BROADCAST

Measurements for element lengths and wavelength spacing are in feet.

kHz	FORMAT: DIR./ DR.EL./ REFL./	SPACING 0.1,	0.125,	0.15,	0.20	0.25
9500	98.4 / 103.5 / 108.4	10.35	12.9375	15.525	20.7	25.875
9520	98.2 / 103.3 / 108.1	10.33	12.9125	15.495	20.66	25.825
9540	98 / 103.1 / 107.9	10.31	12.8875	15.465	20.62	25.775
9560	97.8 / 102.9 / 107.7	10.29	12.8625	15.435	20.58	25.725
9580	97.5 / 102.7 / 107.5	10.27	12.8375	15.405	20.54	25.675
9600	97.3 / 102.5 / 107.2	10.25	12.8125	15.375	20.5	25.625
9620	97.1 / 102.2 / 107	10.22	12.775	15.33	20.44	25.55
9640	96.9 / 102 / 106.8	10.2	12.75	15.3	20.4	25.5
9660	96.7 / 101.8 / 106.6	10.18	12.725	15.27	20.36	25.45
9680	96.5 / 101.6 / 106.4	10.16	12.7	15.24	20.32	25.4
9700	96.3 / 101.4 / 106.1	10.14	12.675	15.21	20.28	25.35
9720	96.1 / 101.2 / 105.9	10.12	12.65	15.18	20.24	25.3
9740	95.9 / 101 / 105.7	10.1	12.625	15.15	20.2	25.25
9760	95.7 / 100.8 / 105.5	10.08	12.6	15.12	20.16	25.2
9780	95.6 / 100.6 / 105.3	10.06	12.575	15.09	20.12	25.15
9800	95.4 / 100.4 / 105.1	10.04	12.55	15.06	20.08	25.1
9820	95.2 / 100.2 / 104.8	10.02	12.525	15.03	20.04	25.05
9840	95 / 100 / 104.6	10	12.5	15	20	25
9860	94.8 / 99.7 / 104.4	9.97	12.4625	14.955	19.94	24.925
9880	94.6 / 99.5 / 104.2	9.95	12.4375	14.925	19.9	24.875
9900	94.4 / 99.3 / 104	9.93	12.4125	14.895	19.86	24.825
9920	94.2 / 99.1 / 103.8	9.91	12.3875	14.865	19.82	24.775
9940	94 / 98.9 / 103.6	9.89	12.3625	14.835	19.78	24.725
9960	93.8 / 98.7 / 103.4	9.87	12.3375	14.805	19.74	24.675
9980	93.6 / 98.5 / 103.2	9.85	12.3125	14.775	19.7	24.625
10000	93.5 / 98.4 / 103	9.84	12.3	14.76	19.68	24.6

TABLE 8-6
THREE-ELEMENT QUAD & DELTA-LOOP ANTENNAS
11500 - 12000 KILOHERTZ
USE: INT'L. BROADCAST

Measurements for element lengths and wavelength spacing are in feet.

kHz	FORMAT: DIR./ DR.EL./REFL./	SPACING 0.1,	0.125,	0.15,	0.20	0.25
11500	81.3 / 85.5 / 89.5	8.55	10.6875	12.825	17.1	21.375
11520	81.1 / 85.4 / 89.4	8.54	10.675	12.81	17.08	21.35
11540	81 / 85.2 / 89.2	8.52	10.65	12.78	17.04	21.3
11560	80.8 / 85.1 / 89.1	8.51	10.6375	12.765	17.02	21.275
11580	80.7 / 84.9 / 88.9	8.49	10.6125	12.735	16.98	21.225
11600	80.6 / 84.8 / 88.7	8.48	10.6	12.72	16.96	21.2
11620	80.4 / 84.6 / 88.6	8.46	10.575	12.69	16.92	21.15
11640	80.3 / 84.5 / 88.4	8.45	10.5625	12.675	16.9	21.125
11660	80.1 / 84.3 / 88.3	8.43	10.5375	12.645	16.86	21.075
11680	80 / 84.2 / 88.1	8.42	10.525	12.63	16.84	21.05
11700	79.9 / 84.1 / 88	8.41	10.5125	12.615	16.82	21.025
11720	79.7 / 83.9 / 87.8	8.39	10.4875	12.585	16.78	20.975
11740	79.6 / 83.8 / 87.7	8.38	10.475	12.57	16.76	20.95
11760	79.5 / 83.6 / 87.5	8.36	10.45	12.54	16.72	20.9
11780	79.3 / 83.5 / 87.4	8.35	10.4375	12.525	16.7	20.875
11800	79.2 / 83.3 / 87.2	8.33	10.4125	12.495	16.66	20.825
11820	79.1 / 83.2 / 87.1	8.32	10.4	12.48	16.64	20.8
11840	78.9 / 83.1 / 86.9	8.31	10.3875	12.465	16.62	20.775
11860	78.8 / 82.9 / 86.8	8.29	10.3625	12.435	16.58	20.725
11880	78.7 / 82.8 / 86.7	8.28	10.35	12.42	16.56	20.7
11900	78.5 / 82.6 / 86.5	8.26	10.325	12.39	16.52	20.65
11920	78.4 / 82.5 / 86.4	8.25	10.3125	12.375	16.5	20.625
11940	78.3 / 82.4 / 86.2	8.24	10.3	12.36	16.48	20.6
11960	78.1 / 82.2 / 86.1	8.22	10.275	12.33	16.44	20.55
11980	78 / 82.1 / 85.9	8.21	10.2625	12.315	16.42	20.525
12000	77.9 / 82 / 85.8	8.2	10.25	12.3	16.4	20.5

TABLE 8-7
THREE-ELEMENT QUAD & DELTA-LOOP ANTENNAS
14000 - 14450 KILOHERTZ
USE: AMATEUR

Measurements for element lengths and wavelength spacing are in feet.

kHz	FORMAT: DIR./DR.EL./REFL./	SPACING 0.1,	0.125,	0.15,	0.20	0.25
14000	66.7 / 70.2 / 73.5	7.02	8.775	10.53	14.04	17.55
14020	66.6 / 70.1 / 73.4	7.01	8.7625	10.515	14.02	17.525
14040	66.5 / 70 / 73.3	7	8.75	10.5	14	17.5
14060	66.5 / 69.9 / 73.2	6.99	8.7375	10.485	13.98	17.475
14080	66.4 / 69.8 / 73.1	6.98	8.725	10.47	13.96	17.45
14100	66.3 / 69.7 / 73	6.97	8.7125	10.455	13.94	17.425
14120	66.2 / 69.6 / 72.9	6.96	8.7	10.44	13.92	17.4
14140	66.1 / 69.5 / 72.8	6.95	8.6875	10.425	13.9	17.375
14160	66 / 69.4 / 72.7	6.94	8.675	10.41	13.88	17.35
14180	65.9 / 69.3 / 72.6	6.93	8.6625	10.395	13.86	17.325
14200	65.8 / 69.2 / 72.5	6.92	8.65	10.38	13.84	17.3
14220	65.7 / 69.1 / 72.4	6.91	8.6375	10.365	13.82	17.275
14240	65.6 / 69.1 / 72.3	6.91	8.6375	10.365	13.82	17.275
14260	65.5 / 69 / 72.2	6.9	8.625	10.35	13.8	17.25
14280	65.4 / 68.9 / 72.1	6.89	8.6125	10.335	13.78	17.225
14300	65.3 / 68.8 / 72	6.88	8.6	10.32	13.76	17.2
14320	65.2 / 68.7 / 71.9	6.87	8.5875	10.305	13.74	17.175
14340	65.2 / 68.6 / 71.8	6.86	8.575	10.29	13.72	17.15
14360	65.1 / 68.5 / 71.7	6.85	8.5625	10.275	13.7	17.125
14380	65 / 68.4 / 71.6	6.84	8.55	10.26	13.68	17.1
14400	64.9 / 68.3 / 71.5	6.83	8.5375	10.245	13.66	17.075
14420	64.8 / 68.2 / 71.4	6.82	8.525	10.23	13.64	17.05
14440	64.7 / 68.1 / 71.3	6.81	8.5125	10.215	13.62	17.025

TABLE 8-8
THREE-ELEMENT QUAD & DELTA-LOOP ANTENNAS
15500 - 16000 KILOHERTZ
USE: INT'L. BROADCAST

Measurements for element lengths and wavelength spacing are in feet.

kHz	FORMAT: DIR./DR.EL./REFL./	SPACING 0.1,	0.125,	0.15,	0.20	0.25
15500	60.3 / 63.4 / 66.4	6.34	7.925	9.51	12.68	15.85
15520	60.2 / 63.4 / 66.3	6.34	7.925	9.51	12.68	15.85
15540	60.1 / 63.3 / 66.2	6.33	7.9125	9.495	12.66	15.825
15560	60 / 63.2 / 66.1	6.32	7.9	9.48	12.64	15.8
15580	60 / 63.1 / 66.1	6.31	7.8875	9.465	12.62	15.775
15600	59.9 / 63 / 66	6.3	7.875	9.45	12.6	15.75
15620	59.8 / 62.9 / 65.9	6.29	7.8625	9.435	12.58	15.725
15640	59.7 / 62.9 / 65.8	6.29	7.8625	9.435	12.58	15.725
15660	59.7 / 62.8 / 65.7	6.28	7.85	9.42	12.56	15.7
15680	59.6 / 62.7 / 65.6	6.27	7.8375	9.405	12.54	15.675
15700	59.5 / 62.6 / 65.6	6.26	7.825	9.39	12.52	15.65
15720	59.4 / 62.5 / 65.5	6.25	7.8125	9.375	12.5	15.625
15740	59.4 / 62.5 / 65.4	6.25	7.8125	9.375	12.5	15.625
15760	59.3 / 62.4 / 65.3	6.24	7.8	9.36	12.48	15.6
15780	59.2 / 62.3 / 65.2	6.23	7.7875	9.345	12.46	15.575
15800	59.1 / 62.2 / 65.1	6.22	7.775	9.33	12.44	15.55
15820	59.1 / 62.1 / 65.1	6.21	7.7625	9.315	12.42	15.525
15840	59 / 62.1 / 65	6.21	7.7625	9.315	12.42	15.525
15860	58.9 / 62 / 64.9	6.2	7.75	9.3	12.4	15.5
15880	58.8 / 61.9 / 64.8	6.19	7.7375	9.285	12.38	15.475
15900	58.8 / 61.8 / 64.7	6.18	7.725	9.27	12.36	15.45
15920	58.7 / 61.8 / 64.6	6.18	7.725	9.27	12.36	15.45
15940	58.6 / 61.7 / 64.6	6.17	7.7125	9.255	12.34	15.425
15960	58.5 / 61.6 / 64.5	6.16	7.7	9.24	12.32	15.4
15980	58.5 / 61.5 / 64.4	6.15	7.6875	9.225	12.3	15.375
16000	58.4 / 61.5 / 64.3	6.15	7.6875	9.225	12.3	15.375

TABLE 8-9
THREE-ELEMENT QUAD & DELTA-LOOP ANTENNAS
16500 - 17000 KILOHERTZ
USE: INT'L. BROADCAST

Measurements for element lengths and wavelength spacing are in feet.

kHz	FORMAT: DIR./DR.EL./REFL./	SPACING 0.1,	0.125,	0.15,	0.20	0.25
16500	56.6 / 59.6 / 62.4	5.96	7.45	8.94	11.92	14.9
16520	56.5 / 59.5 / 62.3	5.95	7.4375	8.925	11.9	14.875
16540	56.5 / 59.4 / 62.2	5.94	7.425	8.91	11.88	14.85
16560	56.4 / 59.4 / 62.1	5.94	7.425	8.91	11.88	14.85
16580	56.3 / 59.3 / 62.1	5.93	7.4125	8.895	11.86	14.825
16600	56.3 / 59.2 / 62	5.92	7.4	8.88	11.84	14.8
16620	56.2 / 59.2 / 61.9	5.92	7.4	8.88	11.84	14.8
16640	56.1 / 59.1 / 61.8	5.91	7.3875	8.865	11.82	14.775
16660	56.1 / 59 / 61.6	5.9	7.375	8.85	11.8	14.75
16680	56 / 58.9 / 61.7	5.89	7.3625	8.835	11.78	14.725
16700	55.9 / 58.9 / 61.6	5.89	7.3625	8.835	11.78	14.725
16720	55.9 / 58.8 / 61.6	5.88	7.35	8.82	11.76	14.7
16740	55.8 / 58.7 / 61.5	5.87	7.3375	8.805	11.74	14.675
16760	55.7 / 58.7 / 61.4	5.87	7.3375	8.805	11.74	14.675
16780	55.7 / 58.6 / 61.3	5.86	7.325	8.79	11.72	14.65
16800	55.6 / 58.5 / 61.3	5.85	7.3125	8.775	11.7	14.625
16820	55.5 / 58.5 / 61.2	5.85	7.3125	8.775	11.7	14.625
16840	55.5 / 58.4 / 61.1	5.84	7.3	8.76	11.68	14.6
16860	55.4 / 58.3 / 61	5.83	7.2875	8.745	11.66	14.575
16880	55.3 / 58.2 / 61	5.82	7.275	8.73	11.64	14.55
16900	55.3 / 58.2 / 60.9	5.82	7.275	8.73	11.64	14.55
16920	55.2 / 58.1 / 60.8	5.81	7.2625	8.715	11.62	14.525
16940	55.1 / 58 / 60.8	5.8	7.25	8.7	11.6	14.5
16960	55.1 / 58 / 60.7	5.8	7.25	8.7	11.6	14.5
16980	55 / 57.9 / 60.6	5.79	7.2375	8.685	11.58	14.475
17000	55 / 57.8 / 60.5	5.78	7.225	8.67	11.56	14.45

TABLE 8-10
THREE-ELEMENT QUAD & DELTA-LOOP ANTENNAS
18500 - 19000 KILOHERTZ
USE: INT'L. BROADCAST

Measurements for element lengths and wavelength spacing are in feet.

kHz	FORMAT: DIR./DR.EL./REFL./	SPACING 0.1,	0.125,	0.15,	0.20	0.25
18500	50.5 / 53.1 / 55.6	5.31	6.6375	7.965	10.62	13.275
18520	50.4 / 53.1 / 55.6	5.31	6.6375	7.965	10.62	13.275
18540	50.4 / 53 / 55.5	5.3	6.625	7.95	10.6	13.25
18560	50.3 / 53 / 55.4	5.3	6.625	7.95	10.6	13.25
18580	50.3 / 52.9 / 55.4	5.29	6.6125	7.935	10.58	13.225
18600	50.2 / 52.9 / 55.3	5.29	6.6125	7.935	10.58	13.225
18620	50.2 / 52.8 / 55.3	5.28	6.6	7.92	10.56	13.2
18640	50.1 / 52.7 / 55.2	5.27	6.5875	7.905	10.54	13.175
18660	50.1 / 52.7 / 55.1	5.27	6.5875	7.905	10.54	13.175
18680	50 / 52.6 / 55.1	5.26	6.575	7.89	10.52	13.15
18700	50 / 52.6 / 55	5.26	6.575	7.89	10.52	13.15
18720	49.9 / 52.5 / 55	5.25	6.5625	7.875	10.5	13.125
18740	49.8 / 52.5 / 54.9	5.25	6.5625	7.875	10.5	13.125
18760	49.8 / 52.4 / 54.9	5.24	6.55	7.86	10.48	13.1
18780	49.7 / 52.3 / 54.8	5.23	6.5375	7.845	10.46	13.075
18800	49.7 / 52.3 / 54.7	5.23	6.5375	7.845	10.46	13.075
18820	49.6 / 52.2 / 54.7	5.22	6.525	7.83	10.44	13.05
18840	49.6 / 52.2 / 54.6	5.22	6.525	7.83	10.44	13.05
18860	49.5 / 52.1 / 54.6	5.21	6.5125	7.815	10.42	13.025
18880	49.5 / 52.1 / 54.5	5.21	6.5125	7.815	10.42	13.025
18900	49.4 / 52 / 54.4	5.2	6.5	7.8	10.4	13
18920	49.4 / 52 / 54.4	5.2	6.5	7.8	10.4	13
18940	49.3 / 51.9 / 54.3	5.19	6.4875	7.785	10.38	12.975
18960	49.3 / 51.8 / 54.3	5.18	6.475	7.77	10.36	12.95
18980	49.2 / 51.8 / 54.2	5.18	6.475	7.77	10.36	12.95
19000	49.2 / 51.7 / 54.2	5.17	6.4625	7.755	10.34	12.925

TABLE 8-11
THREE-ELEMENT QUAD & DELTA-LOOP ANTENNAS
21000 - 21450 KILOHERTZ
USE: AMATEUR

Measurements for element lengths and wavelength spacing are in feet.

kHz	FORMAT: DIR./DR.EL./REFL./	SPACING 0.1,	0.125,	0.15,	0.20	0.25
21000	44.5 / 46.8 / 49	4.68	5.85	7.02	9.36	11.7
21020	44.4 / 46.8 / 49	4.68	5.85	7.02	9.36	11.7
21040	44.4 / 46.7 / 48.9	4.67	5.8375	7.005	9.34	11.675
21060	44.3 / 46.7 / 48.9	4.67	5.8375	7.005	9.34	11.675
21080	44.3 / 46.6 / 48.8	4.66	5.825	6.99	9.32	11.65
21100	44.3 / 46.6 / 48.8	4.66	5.825	6.99	9.32	11.65
21120	44.2 / 46.5 / 48.7	4.65	5.8125	6.975	9.3	11.625
21140	44.2 / 46.5 / 48.7	4.65	5.8125	6.975	9.3	11.625
21160	44.1 / 46.5 / 48.6	4.65	5.8125	6.975	9.3	11.625
21180	44.1 / 46.4 / 48.6	4.64	5.8	6.96	9.28	11.6
21200	44.1 / 46.4 / 48.5	4.64	5.8	6.96	9.28	11.6
21220	44 / 46.3 / 48.5	4.63	5.7875	6.945	9.26	11.575
21240	44 / 46.3 / 48.4	4.63	5.7875	6.945	9.26	11.575
21260	43.9 / 46.2 / 48.4	4.62	5.775	6.93	9.24	11.55
21280	43.9 / 46.2 / 48.4	4.62	5.775	6.93	9.24	11.55
21300	43.8 / 46.1 / 48.3	4.61	5.7625	6.915	9.22	11.525
21320	43.8 / 46.1 / 48.3	4.61	5.7625	6.915	9.22	11.525
21340	43.8 / 46.1 / 48.2	4.61	5.7625	6.915	9.22	11.525
21360	43.7 / 46 / 48.2	4.6	5.75	6.9	9.2	11.5
21380	43.7 / 46 / 48.1	4.6	5.75	6.9	9.2	11.5
21400	43.6 / 45.9 / 48.1	4.59	5.7375	6.885	9.18	11.475
21420	43.6 / 45.9 / 48	4.59	5.7375	6.885	9.18	11.475
21440	43.6 / 45.8 / 48	4.58	5.725	6.87	9.16	11.45

TABLE 8-12
THREE-ELEMENT QUAD & DELTA-LOOP ANTENNAS
25500 - 26000 KILOHERTZ
USE: INT'L. BROADCAST

Measurements for element lengths and wavelength spacing are in feet.

kHz	FORMAT: DIR./DR.EL./REFL./	SPACING 0.1,	0.125,	0.15,	0.20	0.25
25500	36.6 / 38.5 / 40.3	3.85	4.8125	5.775	7.7	9.625
25520	36.6 / 38.5 / 40.3	3.85	4.8125	5.775	7.7	9.625
25540	36.6 / 38.5 / 40.3	3.85	4.8125	5.775	7.7	9.625
25560	36.5 / 38.4 / 40.2	3.84	4.8	5.76	7.68	9.6
25580	36.5 / 38.4 / 40.2	3.84	4.8	5.76	7.68	9.6
25600	36.5 / 38.4 / 40.2	3.84	4.8	5.76	7.68	9.6
25620	36.4 / 38.4 / 40.2	3.84	4.8	5.76	7.68	9.6
25640	36.4 / 38.3 / 40.1	3.83	4.7875	5.745	7.66	9.575
25660	36.4 / 38.3 / 40.1	3.83	4.7875	5.745	7.66	9.575
25680	36.4 / 38.3 / 40.1	3.83	4.7875	5.745	7.66	9.575
25700	36.3 / 38.2 / 40	3.82	4.775	5.73	7.64	9.55
25720	36.3 / 38.2 / 40	3.82	4.775	5.73	7.64	9.55
25740	36.3 / 38.2 / 40	3.82	4.775	5.73	7.64	9.55
25760	36.2 / 38.1 / 39.9	3.81	4.7625	5.715	7.62	9.525
25780	36.2 / 38.1 / 39.9	3.81	4.7625	5.715	7.62	9.525
25800	36.2 / 38.1 / 39.9	3.81	4.7625	5.715	7.62	9.525
25820	36.2 / 38.1 / 39.8	3.81	4.7625	5.715	7.62	9.525
25840	36.1 / 38 / 39.8	3.8	4.75	5.7	7.6	9.5
25860	36.1 / 38 / 39.8	3.8	4.75	5.7	7.6	9.5
25880	36.1 / 38 / 39.7	3.8	4.75	5.7	7.6	9.5
25900	36.1 / 37.9 / 39.7	3.79	4.7375	5.685	7.58	9.475
25920	36 / 37.9 / 39.7	3.79	4.7375	5.685	7.58	9.475
25940	36 / 37.9 / 39.7	3.79	4.7375	5.685	7.58	9.475
25960	36 / 37.9 / 39.6	3.79	4.7375	5.685	7.58	9.475
25980	35.9 / 37.8 / 39.6	3.78	4.725	5.67	7.56	9.45
26000	35.9 / 37.8 / 39.6	3.78	4.725	5.67	7.56	9.45

TABLE 8-13
THREE-ELEMENT QUAD & DELTA-LOOP ANTENNAS
26500 - 27000 KILOHERTZ
USE: INT'L. BROADCAST

Measurements for element lengths and wavelength spacing are in feet.

kHz	FORMAT: DIR./DR.EL./REFL./	SPACING 0.1,	0.125,	0.15,	0.20	0.25
26500	35.2 / 37.1 / 38.8	3.71	4.6375	5.565	7.42	9.275
26520	35.2 / 37.1 / 38.8	3.71	4.6375	5.565	7.42	9.275
26540	35.2 / 37 / 38.8	3.7	4.625	5.55	7.4	9.25
26560	35.2 / 37 / 38.7	3.7	4.625	5.55	7.4	9.25
26580	35.1 / 37 / 38.7	3.7	4.625	5.55	7.4	9.25
26600	35.1 / 36.9 / 38.7	3.69	4.6125	5.535	7.38	9.225
26620	35.1 / 36.9 / 38.6	3.69	4.6125	5.535	7.38	9.225
26640	35 / 36.9 / 38.6	3.69	4.6125	5.535	7.38	9.225
26660	35 / 36.9 / 38.6	3.69	4.6125	5.535	7.38	9.225
26680	35 / 36.8 / 38.6	3.68	4.6	5.52	7.36	9.2
26700	35 / 36.8 / 38.5	3.68	4.6	5.52	7.36	9.2
26720	34.9 / 36.8 / 38.5	3.68	4.6	5.52	7.36	9.2
26740	34.9 / 36.7 / 38.5	3.67	4.5875	5.505	7.34	9.175
26760	34.9 / 36.7 / 38.4	3.67	4.5875	5.505	7.34	9.175
26780	34.9 / 36.7 / 38.4	3.67	4.5875	5.505	7.34	9.175
26800	34.8 / 36.7 / 38.4	3.67	4.5875	5.505	7.34	9.175
26820	34.8 / 36.6 / 38.4	3.66	4.575	5.49	7.32	9.15
26840	34.8 / 36.6 / 38.3	3.66	4.575	5.49	7.32	9.15
26860	34.8 / 36.6 / 38.3	3.66	4.575	5.49	7.32	9.15
26880	34.7 / 36.6 / 38.3	3.66	4.575	5.49	7.32	9.15
26900	34.7 / 36.5 / 38.2	3.65	4.5625	5.475	7.3	9.125
26920	34.7 / 36.5 / 38.2	3.65	4.5625	5.475	7.3	9.125
26940	34.7 / 36.5 / 38.2	3.65	4.5625	5.475	7.3	9.125
26960	34.6 / 36.4 / 38.2	3.64	4.55	5.46	7.28	9.1
26980	34.6 / 36.4 / 38.1	3.64	4.55	5.46	7.28	9.1
27000	34.6 / 36.4 / 38.1	3.64	4.55	5.46	7.28	9.1

TABLE 8-14
THREE-ELEMENT QUAD & DELTA-LOOP ANTENNAS
28500 - 29000 KILOHERTZ
USE: AMATEUR

Measurements for element lengths and wavelength spacing are in feet.

kHz	FORMAT: DIR./DR.EL./REFL./	SPACING 0.1,	0.125,	0.15,	0.20	0.25
28500	32.8 / 34.5 / 36.1	3.45	4.3125	5.175	6.9	8.625
28520	32.7 / 34.5 / 36.1	3.45	4.3125	5.175	6.9	8.625
28540	32.7 / 34.4 / 36	3.44	4.3	5.16	6.88	8.6
28560	32.7 / 34.4 / 36	3.44	4.3	5.16	6.88	8.6
28580	32.7 / 34.4 / 36	3.44	4.3	5.16	6.88	8.6
28600	32.6 / 34.4 / 36	3.44	4.3	5.16	6.88	8.6
28620	32.6 / 34.3 / 35.9	3.43	4.2875	5.145	6.86	8.575
28640	32.6 / 34.3 / 35.9	3.43	4.2875	5.145	6.86	8.575
28660	32.6 / 34.3 / 35.9	3.43	4.2875	5.145	6.86	8.575
28680	32.6 / 34.3 / 35.9	3.43	4.2875	5.145	6.86	8.575
28700	32.5 / 34.2 / 35.8	3.42	4.275	5.13	6.84	8.55
28720	32.5 / 34.2 / 35.8	3.42	4.275	5.13	6.84	8.55
28740	32.5 / 34.2 / 35.8	3.42	4.275	5.13	6.84	8.55
28760	32.5 / 34.2 / 35.8	3.42	4.275	5.13	6.84	8.55
28780	32.4 / 34.1 / 35.7	3.41	4.2625	5.115	6.82	8.525
28800	32.4 / 34.1 / 35.7	3.41	4.2625	5.115	6.82	8.525
28820	32.4 / 34.1 / 35.7	3.41	4.2625	5.115	6.82	8.525
28840	32.4 / 34.1 / 35.7	3.41	4.2625	5.115	6.82	8.525
28860	32.3 / 34 / 35.6	3.4	4.25	5.1	6.8	8.5
28880	32.3 / 34 / 35.6	3.4	4.25	5.1	6.8	8.5
28900	32.3 / 34 / 35.6	3.4	4.25	5.1	6.8	8.5
28920	32.3 / 34 / 35.6	3.4	4.25	5.1	6.8	8.5
28940	32.3 / 34 / 35.5	3.4	4.25	5.1	6.8	8.5
28960	32.2 / 33.9 / 35.5	3.39	4.2375	5.085	6.78	8.475
28980	32.2 / 33.9 / 35.5	3.39	4.2375	5.085	6.78	8.475
29000	32.2 / 33.9 / 35.5	3.39	4.2375	5.085	6.78	8.475

TABLE 8-15
THREE-ELEMENT QUAD & DELTA-LOOP ANTENNAS
29000 - 29500 KILOHERTZ
USE: AMATEUR

Measurements for element lengths and wavelength spacing are in feet.

kHz	FORMAT: DIR./DR.EL./REFL./	SPACING 0.1,	0.125,	0.15,	0.20	0.25
29000	32.2 / 33.9 / 35.5	3.39	4.2375	5.085	6.78	8.475
29020	32.2 / 33.9 / 35.4	3.39	4.2375	5.085	6.78	8.475
29040	32.1 / 33.8 / 35.4	3.38	4.225	5.07	6.76	8.45
29060	32.1 / 33.8 / 35.4	3.38	4.225	5.07	6.76	8.45
29080	32.1 / 33.8 / 35.4	3.38	4.225	5.07	6.76	8.45
29100	32.1 / 33.8 / 35.3	3.38	4.225	5.07	6.76	8.45
29120	32.1 / 33.7 / 35.3	3.37	4.2125	5.055	6.74	8.425
29140	32 / 33.7 / 35.3	3.37	4.2125	5.055	6.74	8.425
29160	32 / 33.7 / 35.3	3.37	4.2125	5.055	6.74	8.425
29180	32 / 33.7 / 35.2	3.37	4.2125	5.055	6.74	8.425
29200	32 / 33.6 / 35.2	3.36	4.2	5.04	6.72	8.4
29220	31.9 / 33.6 / 35.2	3.36	4.2	5.04	6.72	8.4
29240	31.9 / 33.6 / 35.2	3.36	4.2	5.04	6.72	8.4
29260	31.9 / 33.6 / 35.2	3.36	4.2	5.04	6.72	8.4
29280	31.9 / 33.6 / 35.1	3.36	4.2	5.04	6.72	8.4
29300	31.9 / 33.5 / 35.1	3.35	4.1875	5.025	6.7	8.375
29320	31.8 / 33.5 / 35.1	3.35	4.1875	5.025	6.7	8.375
29340	31.8 / 33.5 / 35.1	3.35	4.1875	5.025	6.7	8.375
29360	31.8 / 33.5 / 35	3.35	4.1875	5.025	6.7	8.375
29380	31.8 / 33.4 / 35	3.34	4.175	5.01	6.68	8.35
29400	31.8 / 33.4 / 35	3.34	4.175	5.01	6.68	8.35
29420	31.7 / 33.4 / 35	3.34	4.175	5.01	6.68	8.35
29440	31.7 / 33.4 / 34.9	3.34	4.175	5.01	6.68	8.35
29460	31.7 / 33.4 / 34.9	3.34	4.175	5.01	6.68	8.35
29480	31.7 / 33.3 / 34.9	3.33	4.1625	4.995	6.66	8.325
29500	31.6 / 33.3 / 34.9	3.33	4.1625	4.995	6.66	8.325

TABLE 8-16
THREE-ELEMENT QUAD & DELTA-LOOP ANTENNAS
29500 - 30000 KILOHERTZ
USE: AMATEUR

Measurements for element lengths and wavelength spacing are in feet.

kHz	FORMAT: DIR./DR.EL./REFL./	SPACING 0.1,	0.125,	0.15,	0.20	0.25
29500	31.6 / 33.3 / 34.9	3.33	4.1625	4.995	6.66	8.325
29520	31.6 / 33.3 / 34.8	3.33	4.1625	4.995	6.66	8.325
29540	31.6 / 33.3 / 34.8	3.33	4.1625	4.995	6.66	8.325
29560	31.6 / 33.2 / 34.8	3.32	4.15	4.98	6.64	8.3
29580	31.6 / 33.2 / 34.8	3.32	4.15	4.98	6.64	8.3
29600	31.5 / 33.2 / 34.7	3.32	4.15	4.98	6.64	8.3
29620	31.5 / 33.2 / 34.7	3.32	4.15	4.98	6.64	8.3
29640	31.5 / 33.1 / 34.7	3.31	4.1375	4.965	6.62	8.275
29660	31.5 / 33.1 / 34.7	3.31	4.1375	4.965	6.62	8.275
29680	31.5 / 33.1 / 34.7	3.31	4.1375	4.965	6.62	8.275
29700	31.4 / 33.1 / 34.6	3.31	4.1375	4.965	6.62	8.275
29720	31.4 / 33.1 / 34.6	3.31	4.1375	4.965	6.62	8.275
29740	31.4 / 33 / 34.6	3.3	4.125	4.95	6.6	8.25
29760	31.4 / 33 / 34.6	3.3	4.125	4.95	6.6	8.25
29780	31.3 / 33 / 34.5	3.3	4.125	4.95	6.6	8.25
29800	31.3 / 33 / 34.5	3.3	4.125	4.95	6.6	8.25
29820	31.3 / 32.9 / 34.5	3.29	4.1125	4.935	6.58	8.225
29840	31.3 / 32.9 / 34.5	3.29	4.1125	4.935	6.58	8.225
29860	31.3 / 32.9 / 34.4	3.29	4.1125	4.935	6.58	8.225
29880	31.2 / 32.9 / 34.4	3.29	4.1125	4.935	6.58	8.225
29900	31.2 / 32.9 / 34.4	3.29	4.1125	4.935	6.58	8.225
29920	31.2 / 32.8 / 34.4	3.28	4.1	4.92	6.56	8.2
29940	31.2 / 32.8 / 34.4	3.28	4.1	4.92	6.56	8.2
29960	31.2 / 32.8 / 34.3	3.28	4.1	4.92	6.56	8.2
29980	31.1 / 32.8 / 34.3	3.28	4.1	4.92	6.56	8.2
30000	31.1 / 32.8 / 34.3	3.28	4.1	4.92	6.56	8.2

Chapter 9
Delta Antennas

The delta loop, or triangle antenna, is a close cousin to the quad—so close, in fact, that the same equations are used to determine the length of the wire around the loop.

Figure 9-1 shows the construction of the delta loop. Note that only the driven element is shown here, but reflectors and directors at 0.125-wavelength spacings may also be used. The shape of the delta loop is an equilateral triangle, i.e., a triangle with all three sides of equal length. The name "delta loop" comes from the fact that this shape resembles the upper case Greek letter "delta" (Δ).

The single element delta loop antenna is a reasonably close match to coaxial cable. In multi-element versions, however, some matching will be required. This is also true if optimum performance of the single element version is to be achieved.

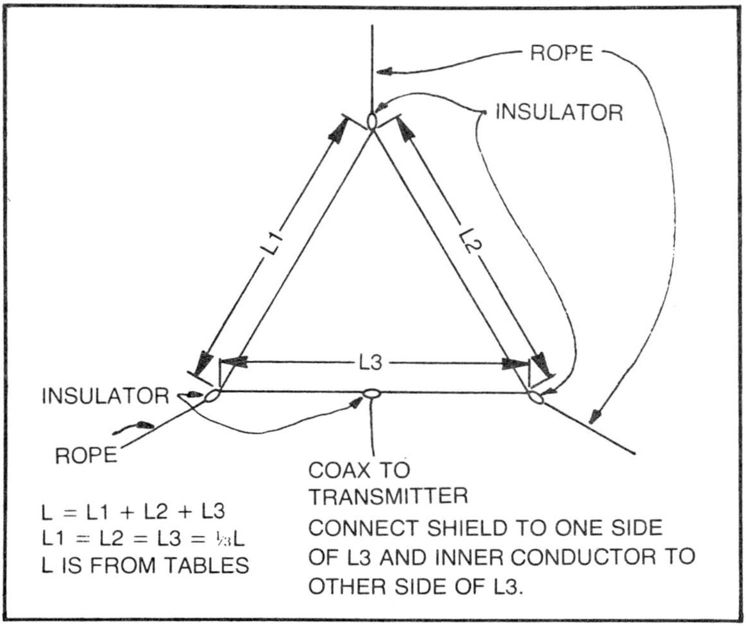

Fig. 9-1. Delta-loop or triangle antenna (one element shown)

To find the overall length of the wire used in the triangle antenna, use either the formulas given in Chapter 8, or the tables for the quad at the end of Chapter 8.

The delta loop is especially favored for low-band DX work. It works well close to the ground. In fact, many amateurs have built the delta loop with the horizontal section only a few feet from the ground! A few even connect the shield from the coaxial cable to the ground, via an underground radial system or a ground rod.

The top point of the delta loop is one of the apexes of the triangle. Although this point is still quite high on 80 meters, it is not impossible. The problems become even less on 40 meters and 20 meters. Once you get to 20 meters and above, however, it might be wise to construct a quad, because it is mechanically easier to deal with, especially with a rotor and some of the commercial construction hardware now on the market.

Chapter 10
CB Antennas

Citizens Band (CB) operators have traditionally used vertically polarized antennas. This is probably due to the fact that CB is primarily a mobile communications activity. The vertically polarized antennas are much easier to install in mobile installations. After all, at 27 MHz, a horizontal dipole would be 18 feet long, and would have to be mounted not less than 10 (preferably 18) feet above the roof of the car in order to be effective. Not very practical. Consequently, CB operators have opted for the vertically polarized antennas.

Even the base station antennas are vertically polarized. It seems that use of cross polarized antennas in a communications system—one vertical and one horizontal antenna—results in loss of effectiveness. There is a lot of loss between the two types. Although nothing is inherently wrong with horizontally polarized antennas on the 27 MHz CB, they do not seem to work quite as well when faced with the fact that all of the mobiles are vertically polarized. It makes just plain common sense, therefore, to use vertically polarized antennas on both ends of the communications circuit.

The types of antennas used by CB-ers include all those employed by amateurs and commercial users. The only requirement is for vertical polarization. In this section, we will discuss the differences between two classes of antennas, the omni-directional and directional, and will hopefully dispel certain rather pernicious myths about antennas.

OMNIDIRECTIONAL VERSUS DIRECTIONAL

There seems to be an almost constant discussion on the merits of various types of antennas, and most of these center around which design is "best." The antenna that is best will depend upon your needs, and the use to which you will put your CB set.

An omnidirectional antenna transmits in all directions. The pattern produced by antennas in this class will be round, or nearly round (when viewed from above). In other words, all directions will "see" the same, or approximately the same, signal strength from an omnidirectional antenna.

Examples of omnidirectional antennas include all of the single element antennas with vertical radiators. This includes quarter-, five-eighths, and three-quarter-wavelength verticals, groundplanes, single element vertical collinears and so forth.

The omnidirectional antenna is used when you want to communicate simultaneously with stations in more than one direction, or when you want to make a general call to anyone. A person seeking help on the CB, for instance, would not want to limit the call to one direction! It might be that the person most able to render assistance rapidly would be in the wrong direction.

Directional antennas, usually called *beams*, favor transmission in only one direction (except for single element horizontal antennas, such as the dipole, which tend to transmit in two directions, and are technically directional, even though bidirectional).

Directional antennas are said to have *power gain*. The antenna does not create power where none existed before! That would be a physical impossibility. No device can create radio frequency power unless power from some other source is consumed. The linear amplifier, of dubious fame, does not create the higher power level "out of the blue," but *converts* power from either batteries, or the 115-volt AC mains, to radio frequency power.

But the beam antenna does produce seemingly greater power. This is a very real, and very useful, advantage of the directional antenna. So how does the directional antenna do this wonderful thing without consuming power from somewhere? It does it by *redirecting* the rf power into only one direction. If you have only a specific amount of rf power to use, and transmit it in all directions—a 360° pattern—then only a small amount is available in any one direction. But, let's suppose that we create a beam antenna that focuses all of the power in one direction, say a 15° beam width (see Fig. 10-1). The total power is now concentrated 360/15 times greater than in the omnidirectional case. This means that it would take 360/15, or 24 times as much power into an omnidirectional antenna to produce the same signal strength at a distant station.

Antenna gain is usually measured in decibels. The correct formula is

$$\text{dB Gain} = 10 \operatorname{Log}_{10} (P1/P2)$$

where dB gain is the gain of the antenna in decibels (power), Log_{10} denotes the use of the common, or base 10 logarithms, P1 is the power produced by the antenna being measured, P2 is the power produced by a reference antenna—normally, a dipole of the same polarity.

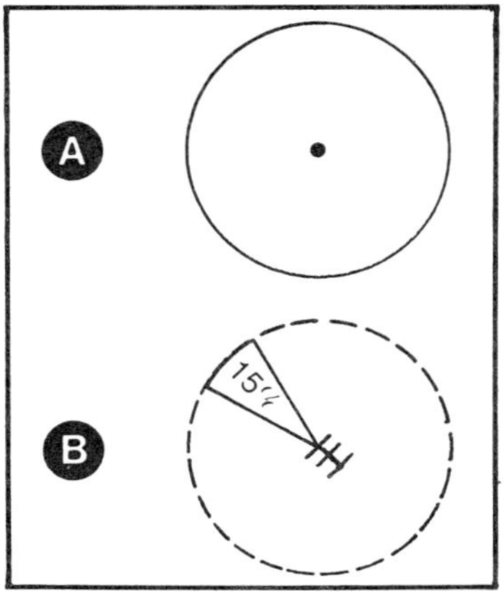

Fig. 10-1. Omnidirectional antenna polar pattern at (A) and beam antenna polar pattern at (B).

Note that we have a reference level, which is the signal intensity produced by a reference antenna. Many of the older antennas used a vertically polarized, half-wavelength, center-fed, dipole antenna as the reference. Recently, however, antenna engineers have used a theoretical concept called the *isotropic radiator* as the reference. This is an ideal point source that expands into a sphere shaped wavefront. The only practical consideration for us is that an antenna measured on the basis of an isotropic radiator will have a gain approximately 2 dB higher than the same measurement taken against a dipole.

How much power am I producting with a gain antenna? The antenna produces an apparent power gain, so many CB-ers are interested in just how much power, in watts, their antenna produces. This measure requires that we know the input power to the antenna and the approximate gain. The power radiated by an antenna is called the *effective radiated power*, or ERP. When a TV or FM station signs off at night, you often hear the announcer say they run "50,000 watts, effective radiated power." This means that they are actually running a lot less *real* power, and use a directional antenna.

To calculate the ERP, use the earlier formula in this chapter, and the information required. Let us work an example in which a CB-er is using a legal transceiver that produces 4 watts of rf output power. Let us assume that the transmission line is very short, so the loss in the coaxial cable is minimal. Let us also assume that the antenna gain is known to be 9 dB.

☐ **Example**

Calculate the ERP of a 9 dB gain antenna if 4 watts is applied at the feedpoint.

Solution:

$$dB = 10 \text{ Log } (P1/P2)$$
$$9 = (10) (\text{Log } (P1/4))$$
$$9/10 = \text{Log } (P1/4)$$

To solve for P1 we must now raise 10 to the power given on each side of the expression, which is easy if you have a scientific calculator.

$$10^{(0.9)} = 10^{(\text{Log } P1/4)}$$
$$7.9 = P1/4$$
$$(4) \times (7.9) = P1$$
$$31.6 \text{ watts} = P1$$

This tells us that 4 watts applied to an antenna with 9 dB gain will produce the same signal strength as almost 32 watts applied to a reference antenna, such as a dipole. Remember the difference between isotropic and dipole references, which would require some adjustment in the 9 dB figure if two different antennas were being compared. The use of gain figures is not so much for system design (at least not on CB—it is important on VHF/UHF and microwave communications). Gain figures are much more useful to compare different types of antenna. The comparison is invalid unless both gain figures are made according to the same rules, using the same type of reference. Note that antenna manufacturers have been known to use a little "creative spec writing" when determining antenna gains. I frankly do not believe fantastic claims of many dB gain. Antenna gain is difficult to obtain with any reliability. I know, for example, few three element beams capable of 15-dB gain, yet I have seen this figure advertised. A table of theoretical gains for the three-element Yagi given in *The Radio Amateur's Handbook (ARRL)* lists the maximum gain as just under 8 dB!

We also see some mixing of two types of gain measurement. There are two types of gain in a beam antenna. The *forward gain* is the gain which we have been worrying about thus far, while the *front-to-back ratio* is the gain of the antenna referenced to *itself.* In the front-to-back ratio measurement, the two power levels produced for the dB gain formula are the power levels measured at a distance from the antenna under two conditions: 1) P1 is when the antenna is pointed toward the measurement site; and 2) P2 is taken when the antenna is pointed away from the measurement site.

But don't get me wrong! There is no problem with the front-to-back ratio, and it is a valid measurement. It does not tell you the same thing as the forward gain, and any manufacturer who uses it in place of the forward gain is, in my opinion, only trying to deceive.

The front-to-back ratio is useful for judging the antenna on *receive*. Does that surprise you—evaluating an antenna on receive? In fact, the most *useful* aspect of the beam antenna on the Citizens Band is its ability to reject stations coming from directions other than the one you want to hear. The same directivity that gives you a bigger *punch* in one direction will reduce the interference from the 10 zillion stations in the other directions! The reduction in interference is very useful. One amateur radio operator, back when rf transistors were still very rare and expensive, contacted a South African station from California, a distance of 16,000 miles, on 96 milliwatts—less power than the unlicensed kids' CB walkie-talkies! The reason why this could *not* be done regularly is the high level of interference on the band. But lower power levels definitely get through when there is less interference. When I first became involved in CB in 1959, we contacted Fredericksburg, VA regularly from Washington, DC, a distance of 50-60 miles. But in those days, though, there was very little interference. Today, it is difficult to work five miles because of all the stations on the air.

There has also been a lot of discussion on the use of multiple radiators in mobile installations. The general consensus among antenna fans is that most CB multiple radiator antennas are at least a waste of money, and at worst will actually reduce the signal level! Note that a properly installed cophased antenna system (installed precisely as to the manufacturer's instructions), will produce an almost omnidirectional pattern. I saw the result of some work produced by a friend of mine, an antenna physicist/engineer, on a NASA antenna test range. They modeled the cophased antenna—a pair of vertical radiators, one-quarter wavelength, spaced one-eighth wavelength apart on the rear trunk lid of a sedan type automobile. The results were that the pattern was almost omnidirectional (at least as good as any other mobile antenna), but the gain was less than the gain of a quarter-wavelength, or 9-foot whip mounted on one of the rear quarter panels. Bumper mounted whips were a different case, as they produced a loss. The gain was, however, better than that expected from the standard 44-inch, loaded, compensation antenna normally used by most CB-ers.

Please note that the 19-inch, heavily loaded antennas, most of which clip on to the roof line of the car, are a disaster gain-wise! They will not produce any gain, and in fact, produce a substantial loss.

It is my opinion that the best mobile compromise for CB antennas is the 44-inch compensation antenna. It is reasonably unobtrusive and works well. Antenna manufacturers, at least those in the CB market, have really gotten their act together over the years to produce an antenna of this sort that works well. If you can do it, of course, use a full length whip. But for most, the best compromise is the 44-inch trunk mounted antenna.

Incidentally, the same research effort that produced these results also found that only a small amount of difference exists between antennas mounted in the center of the trunk lid and the center of the roof. The standard wisdom among CB-ers for years was that the center of the roof was "much, much better." But this is not born out by the empirical evidence from a professional antenna test facility. It is better, but not very much better. The trunk mounted antenna has far fewer problems, and that makes the difference in performance negligible. For example, the trunk mounted antenna is a lot easier to install, it is not as prone to serious accident during installation—a torn headliner, it will not damage the car nearly as much if the base gasket has a leak (the headliner inside the car will become damaged if a roof mounted antenna leaks), and the antenna on the trunk is not nearly as prone to being damaged by car washes and overhanging branches and other obstacles.

COAXIAL CABLE LOSSES

Coaxial cable is both the best and the worst type of antenna transmission line. On the plus side, it is shielded and is easy to install. It will also tolerate a fair amount of abuse and sloppiness in installation (which twinlead will not). This is why really professional TV antenna installers opt for coax, even though twinlead is available to them at lower cost. On the minus side, however, coax is lossy. A long run of coaxial cable will reduce your signal appreciably. The losses increase as frequency increases. In general, the foam dielectric cables are less lossy than regular polyethylene dielectric cables. Also, the larger sizes (RG-8 and RG-11) are less lossy than the smaller types (RG-58 and RG-59).

The general rule is that runs under 30 to 50 feet can accommodate the smaller, easier to work, sizes. But for runs over 50 feet, use the larger size cable. Since CB operators who run legal transmitters have so little power to begin with, it may be wise to always use the larger type of cable on base stations, where the runs tend to be longer, and reserve the small diameter cable for mobile installations, where the larger stuff is very difficult to handle, and the typical antenna will not accommodate it.

TRANSMISSION LINE LENGTHS

Does the length of the transmission line between a CB transmitter and the antenna make any difference? The answer can be both "yes" and "no." As far as the performance of the system, as viewed by a distant station receiving the signal, the absolute length is totally unimportant. Of course, this statement excluded the case where the transmission line is so long that it cuts the signal way down (every dB of loss represents approximately 25 percent loss of power). But when we are making measurements of the antenna system, the length of the cable can be critical.

All antenna measurements involving SWR and impedance (see Chapters 15, 16 and 18) must be made through transmission lines that are *integer multiples of an electrical half wavelength.* Different types of transmission line have different velocity factors. For regular polyethylene dielectric, coaxial cable the velocity factor is 0.66, so the lengths are found from column three (most are about 12 feet). Foam dielectric cables, on the other hand, have a higher velocity factor (0.80), so you would use the lengths give in the fourth column to find electrical half wavelengths.

The SWR readings that you obtain from the typical CB SWR meter are valid only under two circumstances: when measured at the base of the antenna (often impossible or very difficult), and when the transmission line is an integer multiple of an electrical half wavelength. The numbers you obtain when making the measurements under any other conditions are totally invalid. Of course, the purpose of tuning an antenna is to reduce the SWR to minimum. The low-cost meter is still valid for this purpose (minimum will be minimum, regardless of the numbers indicated), but don't believe the numbers obtained. If you tell me that your system has an SWR of 1:1, and don't know how long the antenna feedline was, I won't believe you.

The SWR is *totally* independent of transmission line length. I don't care what your buddy, or the guy at the CB shop tells you, the transmission line length has no effect on real SWR. The pernicious myth that you can cut the coaxial cable length to obtain the best SWR is rooted in an observation using low-cost voltage—or current—operated meters. These meters are valid under only the two conditions given above. At any other point in the line they do not produce believable measurements and will lead you astray by making you think that the line can be trimmed to produce a lower SWR.

SWR is a measure of the mismatch between the antenna and the transmission line. The only adjustment of SWR that is worthwhile is one that improves the match between the antenna and the transmission line. On most antennas, this means adjustment of the *length of the antenna,* not the length of the transmission line. In some other cases—beam and quad antennas using gamma or T-matches—it is adjustment of the matching device that produces best SWR.

Do not ever bother adjusting the transmission line to obtain the best SWR. You are wasting time and effort. Adjust the antenna length or the matching network to obtain the lowest SWR. Try to always cut the transmission line length to an integer (i.e. 1, 2, 3 ...) multiple of an electrical half wavelength.

TABLE 10-1
LENGTHS FOR BEAM ANTENNAS ON C.B.

Channel	MHz	YAGI D/D.E./R	QUAD & DELTA D/D.E./R	0.25,	0.5,	0.1,	0.125	0.15,	0.2
1	26.965	16.68/ 17.35/ 18.24	34.67/ 36.49/ 38.19	9.12	18.24	3.64	4.56	5.47	7.29
2	26.975	16.68/ 17.34/ 18.23	34.66/ 36.47/ 38.18	9.11	18.23	3.64	4.55	5.47	7.29
3	26.985	16.67/ 17.34/ 18.23	34.64/ 36.46/ 38.16	9.11	18.23	3.64	4.55	5.46	7.29
4	27.005	16.66/ 17.33/ 18.21	34.62/ 36.43/ 38.14	9.1	18.21	3.64	4.55	5.46	7.28
5	27.015	16.65/ 17.32/ 18.21	34.61/ 36.42/ 38.12	9.1	18.21	3.64	4.55	5.46	7.28
6	27.025	16.65/ 17.31/ 18.2	34.59/ 36.41/ 38.11	9.1	18.2	3.64	4.55	5.46	7.28
7	27.035	16.64/ 17.31/ 18.19	34.58/ 36.39/ 38.09	9.09	18.19	3.63	4.54	5.45	7.27
8	27.055	16.63/ 17.29/ 18.18	34.55/ 36.37/ 38.07	9.09	18.18	3.63	4.54	5.45	7.27
9	27.065	16.62/ 17.29/ 18.17	34.54/ 36.35/ 38.05	9.08	18.17	3.63	4.54	5.45	7.27
10	27.075	16.62/ 17.28/ 18.17	34.53/ 36.34/ 38.04	9.08	18.17	3.63	4.54	5.45	7.26
11	27.085	16.61/ 17.27/ 18.16	34.52/ 36.33/ 38.02	9.08	18.16	3.63	4.54	5.44	7.26
12	27.105	16.6/ 17.26/ 18.15	34.49/ 36.3/ 38	9.07	18.15	3.63	4.53	5.44	7.26
13	27.115	16.59/ 17.25/ 18.14	34.48/ 36.28/ 37.98	9.07	18.14	3.62	4.53	5.44	7.25
14	27.125	16.58/ 17.25/ 18.13	34.47/ 36.27/ 37.97	9.06	18.13	3.62	4.53	5.44	7.25
15	27.135	16.58/ 17.24/ 18.13	34.45/ 36.26/ 37.95	9.06	18.13	3.62	4.53	5.43	7.25
16	27.155	16.57/ 17.23/ 18.11	34.43/ 36.23/ 37.93	9.05	18.11	3.62	4.52	5.43	7.24
17	27.165	16.56/ 17.22/ 18.11	34.41/ 36.22/ 37.91	9.05	18.11	3.62	4.52	5.43	7.24
18	27.175	16.55/ 17.22/ 18.1	34.4/ 36.2/ 37.9	9.05	18.1	3.62	4.52	5.43	7.24
19	27.185	16.55/ 17.21/ 18.09	34.39/ 36.19/ 37.88	9.04	18.09	3.61	4.52	5.42	7.23
20	27.205	16.54/ 17.2/ 18.08	34.36/ 36.16/ 37.86	9.04	18.08	3.61	4.52	5.42	7.23
21	27.215	16.53/ 17.19/ 18.07	34.35/ 36.15/ 37.84	9.03	18.07	3.61	4.51	5.42	7.23
22	27.225	16.52/ 17.19/ 18.07	34.34/ 36.14/ 37.83	9.03	18.07	3.61	4.51	5.42	7.22
23	27.255	16.51/ 17.17/ 18.05	34.3/ 36.1/ 37.79	9.02	18.05	3.61	4.51	5.41	7.22
24	27.235	16.52/ 17.18/ 18.06	34.33/ 36.12/ 37.81	9.03	18.06	3.61	4.51	5.41	7.22
25	27.245	16.51/ 17.17/ 18.05	34.31/ 36.11/ 37.8	9.02	18.05	3.61	4.51	5.41	7.22
26	27.265	16.5/ 17.16/ 18.04	34.29/ 36.09/ 37.77	9.02	18.04	3.6	4.51	5.41	7.21
27	27.275	16.49/ 17.15/ 18.03	34.28/ 36.07/ 37.76	9.01	18.03	3.6	4.5	5.41	7.21
28	27.285	16.49/ 17.15/ 18.03	34.26/ 36.06/ 37.74	9.01	18.03	3.6	4.5	5.4	7.21
29	27.295	16.48/ 17.14/ 18.02	34.25/ 36.05/ 37.73	9.01	18.02	3.6	4.5	5.4	7.21
30	27.305	16.48/ 17.13/ 18.01	34.24/ 36.03/ 37.72	9	18.01	3.6	4.5	5.4	7.2
31	27.315	16.47/ 17.13/ 18.01	34.23/ 36.02/ 37.7	9	18.01	3.6	4.5	5.4	7.2
32	27.325	16.46/ 17.12/ 18	34.21/ 36.01/ 37.69	9	18	3.6	4.5	5.4	7.2
33	27.335	16.46/ 17.12/ 17.99	34.2/ 35.99/ 37.68	8.99	17.99	3.59	4.49	5.39	7.19
34	27.345	16.45/ 17.11/ 17.99	34.19/ 35.98/ 37.66	8.99	17.99	3.59	4.49	5.39	7.19
35	27.355	16.45/ 17.1/ 17.98	34.18/ 35.97/ 37.65	8.99	17.98	3.59	4.49	5.39	7.19
36	27.365	16.44/ 17.1/ 17.97	34.16/ 35.95/ 37.63	8.98	17.97	3.59	4.49	5.39	7.19
37	27.375	16.43/ 17.09/ 17.97	34.15/ 35.94/ 37.62	8.98	17.97	3.59	4.49	5.39	7.18
38	27.385	16.43/ 17.08/ 17.96	34.14/ 35.93/ 37.61	8.98	17.96	3.59	4.49	5.38	7.18
39	27.395	16.42/ 17.08/ 17.95	34.13/ 35.91/ 37.59	8.97	17.95	3.59	4.48	5.38	7.18
40	27.405	16.42/ 17.07/ 17.95	34.11/ 35.9/ 37.58	8.97	17.95	3.59	4.48	5.38	7.18

Chapter 11
FM Broadcast Antennas

Ordinarily, one does not want any form of directional antenna on a broadcast band. The stations will inevitably be located in different directions, so directivity on an antenna makes it necessary to rotate the antenna towards each new station. But in the FM broadcast band, 88 to 108 MHz, there may well be good reasons why a directional antenna is needed. For one thing, there are multipath reflections at these frequencies that do not occur at AM frequencies. The multipath phenomenon is also the cause of ghosts on television reception.

The most common form of FM antenna is a half-wavelength folded dipole (see Chapter 1) made from 300-ohm TV-type twin-lead transmission line. Table 11-1 gives you the correct overall length of the folded dipole in inches (use the half-wavelength values). The "radiator" best suited to covering the whole band is approximately 56 inches long, but you may wish to cut for a specific frequency in order to receive one station best.

Cut the twinlead to a length of 56 inches, or to the length obtained from Table 11-1. Strip ¼-inch of insulation from each conductor, on each end, and tie the ends together per Fig. 11-1. Solder the two conductors together on each end of the cable. Next, cut one of the conductors of the twin-lead in the exact center (again, see Fig. 11-1). Solder another piece of twinlead to these two conductors per the inset in the figure. This new twinlead is the 300-ohm transmission line to the receiver.

This antenna should be mounted as high as possible, but will improve the performance of many stereo receivers even when close to the floor. In fact, many manufacturers of console stereo sets tape an FM folded dipole to the inside of the cabinet, and it works fine for most customers. Small FM radios often have an auxilliary FM antenna input, but normally use the AC power mains cord as the FM antenna. A small value capacitor connects the input of the FM receiver to the line cord. But this is poor at best, and only works well in very high signal strength areas. These sets can be improved substantially by the use of the folded dipole.

Table 11-1. Overall length of folded dipole in inches.

ANTENNA DIMENSIONS FOR FM BROADCAST BAND 88 TO 108 MHz ALL DIMENSIONS ARE INCHES					
FREQ.	¼-λ	½-λ	FREQ.	¼-λ	½-λ
88.1	31.9	63.7	98.1	28.6	57.2
88.5	31.7	63.5	98.5	28.5	57.0
88.9	31.6	63.2	98.9	28.4	56.8
89.3	31.4	62.9	99.3	28.3	56.6
89.7	31.3	62.6	99.7	28.2	56.3
90.1	31.2	62.3	100.1	28.1	56.1
90.5	31.0	62.1	100.5	27.9	55.9
90.9	30.9	61.8	100.9	27.8	55.7
91.3	30.8	61.5	101.3	27.7	55.4
91.7	30.6	61.2	101.7	27.6	55.2
92.1	30.5	61.0	102.1	27.5	55.0
92.5	30.4	60.7	102.5	27.4	54.8
92.9	30.2	60.5	102.9	27.3	54.6
93.3	30.1	60.2	103.3	27.2	54.4
93.7	30.0	59.9	103.7	27.1	54.2
94.1	29.8	59.7	104.1	27.0	53.9
94.5	29.7	59.4	104.5	26.9	53.7
94.9	29.6	59.2	104.9	26.8	53.5
95.3	29.5	58.9	105.3	26.7	53.3
95.7	29.3	58.7	105.7	26.6	53.1
96.1	29.2	58.4	106.1	26.5	52.9
96.5	29.1	58.2	106.5	26.4	52.7
96.9	29.0	58.0	106.9	26.3	52.5
97.3	28.9	57.7	107.3	26.2	52.3
97.7	28.7	57.5	107.7	26.1	52.1

If you mount this antenna to the wall or ceiling, do not use thumb tacks! Although it is possible to "get away with it," these deteriorate the performance of the antenna. If you must use a metal fastener, use a staple inserted into the insulation of the twinlead along the axis of the conductors. Place it as close to the center of the space between the two conductors as possible. The best fastener, though, is anything nonmetallic. The traditional solution was tape, but now you can buy adhesive-backed picture hangers that work wonders!

If you have an FM broadcast receiver in your care, then the car antenna may well be a vertical. The proper length for the mast is

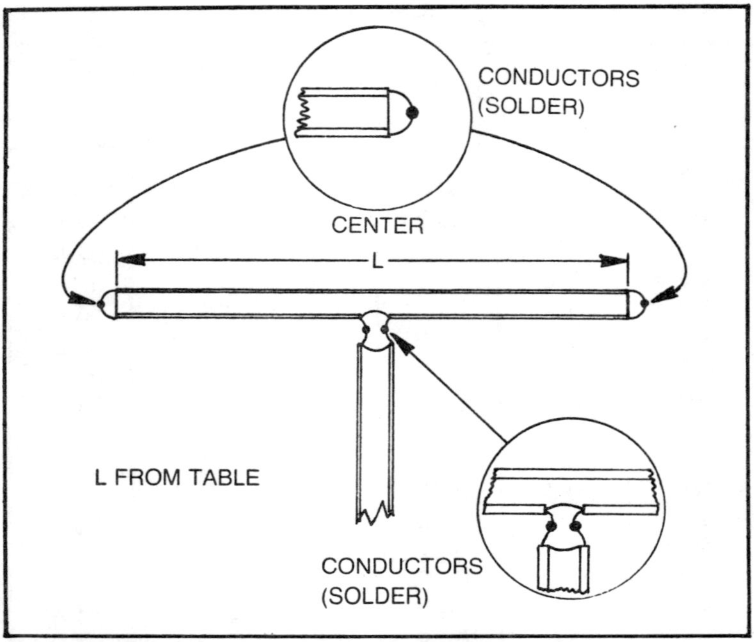

Fig. 11-1. Practical construction of a folded dipole for the FM broadcast band.

about 31 inches, although you can use the quarter-wavelength figures from Table 11-1 to bias the antenna toward one frequency or another. If the mast is adjustable, you will note a substantial improvement—most of the time—when the height of the mast is proper for the frequency being received.

Chapter 12

Antenna Calculation Programs

Microcomputers are now part of the electronic hobby scene. Where computers were once expensive and exotic beyond measure, as far as the individual was concerned, they can now be considered almost tiresomely common place. Many local amateur radio clubs, especially those tending to attract technically oriented people rather than operators, will have a few amateur computerniks among the hams. In this chapter we will present the programs used to create the tables given in the various chapters. Also presented, although in Chapter 14, is a program for calculating the bearings from true north, and the distance in arc length, miles or kilometers, between your QTH and any station, city, or country on earth. If you have ever priced the antenna bearings charts offered by some computer services, you will know that this program, alone, pays half the cost of the book!

Most of the programs are written in Apple II BASIC (not the mini-BASIC). Care must be taken when translating them to any other basic to make sure the syntax is correct, and that the command called for is, indeed, available. If, for example, your version of BASIC uses the letters SQR for square root, then these programs should run as written. But if your BASIC uses the letters SQRT, then modify the listing to suit your computer. Similarly, in the bearings program of the other chapter, we used the ATN (arc tangent) function, and this is not always available in all versions of BASIC. If you lack ATN (or whatever it is called), then see Chapter 13 first.

Programs numbers 7 and 8 were written in Digital Group's MAXI-BASIC. This is not in itself too important, except that MAXI-BASIC allows the use of print out format commands. In the output print statements you might see something like %6F2. This means that the output format will appear in a six character field (of which the decimal point is one character), with two characters to the right of the decimal point. This means that an answer like "123.127895" in most versions of BASIC would print out only the six left-most characters, including the decimal point, such as 123.12. If your version of BASIC does not allow formatted output, then simply delete the format commands and be content with the zillion-place answers.

PROGRAM NUMBER 1

This program—and similar versions—was used to generate the charts for the single element antennas, i.e. dipoles, verticals, etc. Lines 5 and 6 were used for housekeeping, along with lines 400 and 1510, so that the tables would be numbered as to chapter and order. You should either modify this to your own use, or delete those lines. The result would be not to print "Table Z-Y" on the top of each chart.

This particular version of the program was used to create the ⅝-wavelengths charts. To modify for other lengths, it is necessary to change two lines, line 9 and 402. Line 9 is the velocity constant needed for calculating ⅝-wavelength in feet.

The other constants are:

1/4-W	234000
1/2-W	468000
3/4-W	702000

Line 402 merely prints out a statement at the top of each chart as to which antenna is being considered. In this case, the print out will be "⅝-WAVELENGTH VERTICAL."

```
 4 LET R = 0
 5 LET Z = 5
 6 LET Y = 1
 9 LET K1 = 585000
10 LET F1 = 3500
11 LET F2 = 4000
12 LET X = 1
13 GOTO 400
15 LET F1 = 5500
16 LET F2 = 6000
17 LET X = 2
18 GOTO 400
20 LET F1 = 6000
21 LET F2 = 6500
22 LET X = 2
23 GOTO 400
25 LET F1 = 7000
26 LET F2 = 7300
27 LET X = 1
28 GOTO 400
30 LET F1 = 9500
31 LET F2 = 10000
32 LET X = 2
33 GOTO 400
35 LET F1 = 11500
```

```
36   LET F2 = 12000
37   LET X = 2
38   GOTO 400
40   LET F1 = 14000
41   LET F2 = 14450
42   LET X = 1
43   GOTO 400
45   LET F1 = 15500
46   LET F2 = 16000
47   LET X = 2
48   GOTO 400
50   LET F1 = 16500
51   LET F2 = 17000
52   LET X = 2
53   GOTO 400
55   LET F1 = 18500
56   LET F2 = 19000
57   LET X = 2
58   GOTO 400
60   LET F1 = 21000
61   LET F2 = 21450
62   LET X = 1
63   GOTO 400
65   LET F1 = 25500
66   LET F2 = 26000
67   LET X = 2
68   GOTO 400
70   LET F1 = 26500
71   LET F2 = 27000
72   LET X = 2
73   GOTO 400
75   LET F1 = 28500
76   LET F2 = 29000
77   LET X = 1
78   GOTO 400
80   LET F1 = 29000
81   LET F2 = 29500
82   LET X = 1
83   GOTO 400
85   LET F1 = 29500
86   LET F2 = 30000
87   LET X = 1
88   GOTO 400
400  PRINT "TABLE ";Z;"-";Y
402  PRINT "5/8-WAVELENGTH VERTICAL"
405  PRINT F1;" - ";F2;" KILOHERTZ"
410  IF X = 1 THEN  GOTO 430
415  IF X = 2 THEN  GOTO 440
430  PRINT "USE: AMATEUR"
432  GOTO 450
440  PRINT "USE: INT'L. BROADCAST"
450  PRINT
451  PRINT
```

```
452  PRINT
500  FOR Q = 1 TO 10
510  LET F(Q) = F1
520  LET L(Q) = K1 / F(Q)
525  LET L(Q) = L(Q) * 100
528  LET L(Q) =  INT (L(Q))
530  LET L(Q) = L(Q) / 100
535  LET A = L(1)
540  LET B = L(2)
545  LET C = L(3)
550  LET D = L(4)
555  LET E = L(5)
560  LET F = L(6)
565  LET G = L(7)
570  LET H = L(8)
575  LET I = L(9)
580  LET J = L(10)
590  LET F1 = F1 + 2
595  NEXT Q
1000 PRINT F(1);"  ";A;"  ";B;"  ";C;"  ";D;"  ";E;"  ";F;"  ";G;"
                    ";H;"  ";I;"  ";J
1050 IF F1 > F2 THEN  GOTO 1500
1100 IF F1 = F2 THEN  GOTO 500
1150 IF F1 < F2 THEN  GOTO 500
1500 LET R = R + 1
1510 LET Y = Y + 1
1520 PRINT
1525 PRINT
1530 PRINT
1570 PRINT "?"
1575 INPUT A
1577 IF A = 2 THEN  GOTO 10000
1600 ON R GOTO 15,20,25,30,35,40,45,50,55,60,65,70,75,80,85,90,95
10000 END
```

PROGRAM NUMBER 2

This is essentially the same as program number 1, except that we consider the three lengths of the director, reflector, and driven element of quad and delta-loop beam antennas. Again, to modify the program, such as for Yagi beam antennas, change the constants and the print out at line 402. In this program, the constants are in lines 1, 2 and 3. For the Yagi antenna these should be:

```
1    LET K1 = 492000
2    LET K2 = 468000
3    LET K3 = 450000
```

The element spacing can be determined from the tables in Chapter 15, and will be 0.1 to 0.25 wavelengths for Yagis, and (most of the time) 0.125 wavelengths for quads and delta loops.

```
LIST

 1  LET K1 = 1030000
 2  LET K2 = 984000
 3  LET K3 = 935000
 4  LET P = 0
 5  LET Z = 8
 6  LET Y = 1
10  LET F1 = 3500
11  LET F2 = 4000
12  LET X = 1
13  GOTO 400
15  LET F1 = 5500
16  LET F2 = 6000
17  LET X = 2
18  GOTO 400
20  LET F1 = 6000
21  LET F2 = 6500
22  LET X = 2
23  GOTO 400
25  LET F1 = 7000
26  LET F2 = 7300
27  LET X = 1
28  GOTO 400
30  LET F1 = 9500
31  LET F2 = 10000
32  LET X = 2
33  GOTO 400
35  LET F1 = 11500
36  LET F2 = 12000
37  LET X = 2
38  GOTO 400
40  LET F1 = 14000
41  LET F2 = 14450
42  LET X = 1
43  GOTO 400
45  LET F1 = 15500
46  LET F2 = 16000
47  LET X = 2
48  GOTO 400
50  LET F1 = 16500
51  LET F2 = 17000
52  LET X = 2
53  GOTO 400
55  LET F1 = 18500
56  LET F2 = 19000
57  LET X = 2
58  GOTO 400
60  LET F1 = 21000
61  LET F2 = 21450
62  LET X = 1
63  GOTO 400
65  LET F1 = 25500
66  LET F2 = 26000
67  LET X = 2
68  GOTO 400
70  LET F1 = 26500
71  LET F2 = 27000
72  LET X = 2
73  GOTO 400
75  LET F1 = 28500
```

```
76  LET F2 = 29000
77  LET X = 1
78  GOTO 400
80  LET F1 = 29000
81  LET F2 = 29500
82  LET X = 1
83  GOTO 400
85  LET F1 = 29500
86  LET F2 = 30000
87  LET X = 1
88  GOTO 400
400  PRINT "TABLE ";Z;"-";Y
402  PRINT "THREE-ELEMENT QUAD & DELTA-LOOP ANTENNAS"
405  PRINT F1;" - ";F2;" KILOHERTZ"
410  IF X = 1 THEN  GOTO 430
415  IF X = 2 THEN  GOTO 440
430  PRINT "USE: AMATEUR"
432  GOTO 450
440  PRINT "USE: INT'L. BROADCAST"
450  PRINT
451  PRINT
452  PRINT "FORMAT: DIR./DR.EL./REFL./ SPACING 0.1, 0.125, 0.15, 0.20 0.25"
453  PRINT
454  PRINT
500  LET D = K3 / F1
510  LET L = K2 / F1
515  LET R = K1 / F1
525  LET D = D * 10
526  LET L = L * 10
527  LET R = R * 10
530  LET D =  INT (D)
531  LET L =  INT (L)
532  LET R =  INT (R)
535  LET D = D / 10
536  LET L = L / 10
537  LET R = R / 10
585  LET A = L * 0.1
586  LET B = L * 0.125
587  LET C = L * 0.15
588  LET U = L * 0.2
589  LET E = L * 0.25
1000  PRINT F1,D;" / ";L;" / ";R,"   ";A,B,C,U,E
1010  LET F1 = F1 + 20
1050  IF F1 > F2 THEN  GOTO 1500
1100  IF F1 = F2 THEN  GOTO 500
1150  IF F1 < F2 THEN  GOTO 500
1500  LET P = P + 1
1510  LET Y = Y + 1
1520  PRINT
1525  PRINT
1530  PRINT
1570  PRINT "?"
1575  INPUT A
1577  IF A = 2 THEN  GOTO 10000
1600  ON P GOTO 15,20,25,30,35,40,45,50,55,60,65,70,75,80,85,80,95
10000  END
```

PROGRAM NUMBER 3

Calculation of VSWR from either forward and reflected voltage or power.

```
LIST

 100  REM  PROGRAM TO COMPUTE SWR
 200  REM  FROM FORWARD AND REFLECTED VOLTAGE OR POWER
 220  FOR Q = 1 TO 5
 225  PRINT
 230  NEXT Q
 240  PRINT "WHICH METHOD?"
 245  PRINT "1. VOLTAGE"
 250  PRINT "2. POWER"
 255  PRINT
 260  PRINT
 270  INPUT A
 280  ON A GOTO 400,1000
 400  PRINT "ENTER FORWARD VOLTAGE"
 410  INPUT E1
 420  PRINT "ENTER REVERSE VOLTAGE"
 430  INPUT E2
 433  IF E2 > E1 THEN  GOTO 4000
 434  IF E1 = E2 THEN  GOTO 5000
 440  LET S = ((E1 + E2) / (E1 - E2))
 450  LET S = S * 100
 455  LET S =  INT (S)
 460  LET S = S / 100
 470  PRINT "VSWR = ";S;":1"
 480  GOTO 2000
1000  PRINT "ENTER FORWARD POWER"
1010  INPUT P1
1020  PRINT "ENTER REFLECTED POWER"
1030  INPUT P2
1035  IF P2 > P1 THEN  GOTO 3000
1036  IF P1 = P2 GOTO 5000
1040  LET P =  SQR (P2 / P1)
1050  LET S1 = 1 + P
1060  LET S2 = 1 - P
1070  LET S = S1 / S2
1080  LET S = S * 100
1090  LET S =  INT (S)
1100  LET S = S / 100
1110  PRINT "VSWR = ";S;":1"
1150  GOTO 2000
2000  PRINT "CONTINUE? PRESS 1  FINISHED? PRESS 2"
2010  INPUT F
2020  ON F GOTO 240,10000
3000  PRINT "YOU GOTTA BE KIDDING!"
3010  PRINT
3020  PRINT "TRY AGAIN WITH A CORRECT VALUE"
```

```
4000  PRINT "YOU GOTTA BE KIDDING!"
4010  PRINT "TRY AGAIN WITH A CORRECT VALUE"
4020  GOTO 400
5000  PRINT "POW! YOUR TRANSMITTER JUST BLEW UP!"
5005  PRINT
5010  GOTO 240
10000  END

]RUN

WHICH METHOD?
1. VOLTAGE
2. POWER

?1
ENTER FORWARD VOLTAGE
?1.66
ENTER REVERSE VOLTAGE
?.06
VSWR = 1.07:1
CONTINUE? PRESS 1  FINISHED? PRESS 2
?1
WHICH METHOD?
1. VOLTAGE
2. POWER

?2
ENTER FORWARD POWER
?200
ENTER REFLECTED POWER
?12.5
VSWR = 1.66:1
CONTINUE? PRESS 1  FINISHED? PRESS 2
?2
```

PROGRAM NUMBER 4

SWR meter calibration program is a program that allows the user to enter upper and lower power limits, and an incremental power. The program will then calcualte the SWR, at the power increments entered, for all values of reflected power between the two entered limits. Lines 270–280 are used to round off the readout, on the theory that an SWR of 1.2243046778:1 is meaningless. A sample run of the program is given, in which power limits of 0 and 10000 watts were entered, and increments of 10 watts were calculated. This data, a means for measuring rf power in the reflected direction (i.e. a Model 43 Thruline® watt meter) will permit you to calibrate any non-power based SWR bridge or meter.

```
LIST

  100  REM  SWR METER CALIBRATION PROGRAM
  110  REM  USING FORWARD AND REFLECTED POWER
  200  PRINT "ENTER LOWEST POSSIBLE REFLECTED POWER"
  210  INPUT P2
  220  PRINT "ENTER MAXIMUM FORWARD POWER"
  225  INPUT P1
  226  PRINT "ENTER POWER INCREMENT PER CALCULATION"
  227  INPUT P3
  229  PRINT "REFL.PWR","VSWR"
  230  LET R =  SQR (P2 / P1)
  240  LET S1 = 1 + R
  250  LET S2 = 1 - R
  260  LET S = S1 / S2
  270  LET S = S * 1000
  275  LET S =  INT (S)
  280  LET S = S / 1000
  290  PRINT P2,S;":1"
  300  LET P2 = P2 + P3
  310  IF P1 = P2 THEN  GOTO 1000
  320  IF P2 < P1 THEN  GOTO 230
  330  IF P2 > P1 THEN  GOTO 1000
 1000  FOR Q = 1 TO 5
 1010  PRINT "THANK YOU"
 1020  NEXT Q
10000   END

 RUN
ENTER LOWEST POSSIBLE REFLECTED POWER
?0
ENTER MAXIMUM FORWARD POWER
?1000
ENTER POWER INCREMENT PER CALCULATION
?10
REFL.PWR       VSWR
0              1:1
10             1.222:1
20             1.329:1
30             1.418:1
40             1.5:1
50             1.576:1
60             1.648:1
70             1.719:1
80             1.788:1
90             1.857:1
100            1.924:1
110            1.992:1
120            2.06:1
130            2.127:1
140            2.195:1
150            2.264:1
160            2.333:1
170            2.403:1
180            2.473:1
190            2.545:1
200            2.618:1
210            2.691:1
220            2.766:1
230            2.843:1
240            2.92:1
```

250	3:1
260	3.08:1
270	3.163:1
280	3.247:1
290	3.333:1
300	3.422:1
310	3.512:1
320	3.604:1
330	3.699:1
340	3.797:1
350	3.897:1
360	4:1
370	4.105:1
380	4.214:1
390	4.326:1
400	4.441:1
410	4.56:1
420	4.683:1
430	4.809:1
440	4.94:1
450	5.075:1
460	5.215:1
470	5.36:1
480	5.51:1
490	5.666:1
500	5.828:1
510	5.996:1
520	6.171:1
530	6.353:1
540	6.542:1
550	6.74:1
560	6.946:1
570	7.162:1
580	7.388:1
590	7.624:1
600	7.872:1
610	8.133:1
620	8.407:1
630	8.695:1
640	9:1
650	9.321:1
660	9.661:1
670	10.021:1
680	10.403:1
690	10.81:1
700	11.244:1
710	11.707:1
720	12.203:1
730	12.736:1
740	13.309:1
750	13.928:1
760	14.598:1
770	15.326:1
780	16.119:1
790	16.988:1
800	17.944:1
810	19:1
820	20.172:1
830	21.482:1
840	22.956:1

```
850             24.626:1
860             26.533:1
870             28.734:1
880             31.301:1
890             34.334:1
900             37.973:1
910             42.42:1
920             47.979:1
930             55.124:1
940             64.651:1
950             77.987:1
960             97.989:1
970             131.325:1
980             197.994:1
990             397.997:1
THANK YOU
```

PROGRAM NUMBER 5

This program converts the fractional feet obtained in most of the other programs to inches. You may use this program as a subroutine to modify the print out of the other programs.

```
 100  REM  PROGRAM TO CONVERT FRACTRIONAL FEET INTO INCHES
 105  REM
 110  REM
 200  PRINT "ENTER FRACTIONAL FEET VALUE"
 210  INPUT F
 220  LET I = F * 12
 300  PRINT F;" FT. = ";I;" IN."
1000  END
```

PROGRAM NUMBER 6

Similar to program number 1, except that the frequencies are all CB channels, this particular version was used to calculate the lengths for 0.25-, 0.5-, 0.625- and 0.75- wavelength vertical radiators. To convert the program to calculate beam antenna lengths, then change the print statements in 142 and 150, and the constants in 210, 220, 230 and 240. In the case of the Yagi, use 450, 468 and 492 for A, B and C, respectively. But for the quad/delta loop use 934, 984 and 1030 for A, B and C, respectively. In the case of the Yagis, let D be the spacing, which will be

SPACING (WAVELENGTH)	D
0.1	98.4
0.125	121
0.15	148
0.20	197
0.25	246

The print out format will then be frequency/ / 0.25-W/ director/driven element/reflector/spacing. Modify the print statement of line 500 if you want another format.

```
LIST

 50  DIM F(40)
101  F(1) = 26.965
102  F(2) = 26.975
103  F(3) = 26.985
104  F(4) = 27.005
105  F(5) = 27.015
106  F(6) = 27.025
107  F(7) = 27.035
108  F(8) = 27.055
109  F(9) = 27.065
110  F(10) = 27.075
111  F(11) = 27.085
112  F(12) = 27.105
113  F(13) = 27.115
114  F(14) = 27.125
115  F(15) = 27.135
116  F(16) = 27.155
117  F(17) = 27.165
118  F(18) = 27.175
119  F(19) = 27.185
120  F(20) = 27.205
121  F(21) = 27.215
122  F(22) = 27.225
123  F(23) = 27.255
124  F(24) = 27.235
125  F(25) = 27.245
126  F(26) = 27.265
127  F(27) = 27.275
128  F(28) = 27.285
129  F(29) = 27.295
130  F(30) = 27.305
131  F(31) = 27.315
132  F(32) = 27.325
133  F(33) = 27.335
134  F(34) = 27.345
135  F(35) = 27.355
136  F(36) = 27.365
137  F(37) = 27.375
138  F(38) = 27.385
139  F(39) = 27.395
140  F(40) = 27.405
141  PRINT "TABLE 10-4"
142  PRINT "LENGTHS FOR NON-BEAM ANTENNAS ON C.B."
143  PRINT
144  PRINT
150  PRINT "FORMAT: 0.25, 0.5, 0.625, 0.75 WAVELENGTHS"
175  PRINT
200  FOR Q = 1 TO 40
210  LET L2 = 234 / F(Q)
220  LET A = 468 / F(Q)
230  LET B = 585 / F(Q)
240  LET C = 702 / F(Q)
330  LET L2 = L2 * 100
340  LET L2 =  INT (L2)
```

```
  350  LET L2 = L2 / 100
  360  LET A = A * 100
  370  LET A =  INT (A)
  380  LET A = A / 100
  390  LET B = B * 100
  391  LET C = C * 100
  392  LET D = D * 100
  395  LET B =  INT (B)
  396  LET C =  INT (C)
  400  LET B = B / 100
  402  LET C = C / 100
  500  PRINT Q,"  (";F(Q);")",L2,A,B,C
  600  NEXT Q
10000  END
```

PROGRAM NUMBER 7

This program is to calculate all of the lengths of the single radiator antennas, the dipoles, verticals and inverted-vee dipoles. It is written in Digital Group MAXI-BASIC.

```
10 REM PROGRAM TO COMPUTE ANTENNA DIMENSIONS
100 FOR Q = 1 TO 20
110 PRINT
120 NEXT Q
121 PRINT
122 PRINT
130 PRINT "1. ½-WAVELENGTH IN FREE-SPACE"
140 PRINT "2. ¼-WAVELENGTH IN FREE-SPACE"
150 PRINT "3. ½-WAVELENGTH DIPOLE"
160 PRINT "4. ¼-WAVELENGTH VERTICAL"
170 PRINT "5. 5/8-WAVELENGTH VERTICAL"
180 PRINT "6. INVERTED VEE"
200 INPUT A
220 ON A GOTO 500, 520, 540, 560, 580, 590
500 LET C = 492000
505 GOTO 700
520 LET C=246000
525 GOTO 700
540 LET C=468000
545 GOTO 700
560 LET C=234000
565 GOTO 700
580 LET C = 585000
585 GOTO 700
590 LET C = 464000
595 GOTO 700
700 PRINT "ENTER LOWER BAND EDGE IN KILOHERTZ"
720 INPUT F1
730 PRINT "ENTER UPPER BAND EDGE IN KILOHERTZ"
740 INPUT F2
750 PRINT "ENTER KILOHERTZ INCREMENT"
760 INPUT F3
770 LET L = C/F1
780 PRINT F1;' KHZ', ,Z6F2;L;' FEET'
790 LET F1 = F1 + F3
800 IF F1 =>(F2 + 53) THEN GOTO 900
820 GOTO 770
900 PRINT "WHAT NOW?"
905 PRINT "1. MORE OF SAME, DATA COMING"
```

```
910 PRINT "2. LET'S DO ANOTHER ANTENNA'
915 PRINT "3. ALL DONE'
930 INPUT Z
960 ON Z GOTO 700, 100, 2000
2000 FOR M = 1 TO 20
2001 PRINT
2005 NEXT M
2010 PRINT 'BYE BYE'
2015 FOR N = 1 TO 30
2020 NEXT N
3000 END
```

PROGRAM NUMBER 8

This program is an expanded version of program number 7. It will calculate antenna dimensions, convert fractional feet to inches, convert feet to metric units, and calculate the physical lengths required to obtain an electrical half wavelength (at various frequencies) of a transmission line.

```
10 REM PROGRAM TO COMPUTE ANTENNA DIMENSIONS, CONVERT
15 REM FRACTIONAL FEET TO INCHES, CONVERT FEET
20 REM TO METRIC UNITS, AND COMPUTE PROPER TRANSMISSION
25 REM LINE LENGTH FOR ELECTRICAL WAVELENGTHS
100 PRINT "WHICH PROGRAM DO YOU WANT?"
110 PRINT "1.ANTENNA DIMENSIONS"
112 PRINT "2. FRACTIONAL FEET TO INCHES CONVERSION"
113 PRINT "3. FEET TO METRIC CONVERSION"
115 PRINT "4. TRANSMISSION LINE DIMENSIONS"
120 INPUT A
150 ON A GOTO 1500, 300, 600, 900
300 FOR Q = 1 TO 20
305 PRINT
310 NEXT Q
315 PRINT "PROGRAM TO CONVERT FRACTIONAL FEET TO INCHES"
320 FOR P = 1 TO 4
321 PRINT
322 NEXT P
325 PRINT "ENTER FRACTIONAL FEET IN DECIMAL FORM"
326 PRINT
330 PRINT "I.E. ENTER 1/4-FEET AS 0.25"
335 INPUT B
340 LET C = B*12
342 FOR X = 1 TO 4
343 PRINT
344 NEXT X
350 PRINT B;" FEET = ", C;" INCHES"
354 FOR M = 1 TO 5
355 PRINT
356 NEXT M
360 PRINT "CONTINUE? PRESS 1 FINISHED? PRESS 2"
365 INPUT B3
370 ON B3 GOTO 300, 10
600 FOR Q = 1 TO 4
605 PRINT
610 NEXT Q
615 PRINT "PROGRAM TO CONVERT FEET TO METRIC UNITS"
620 FOR P = 1 TO 4
621 PRINT
622 NEXT P
625 PRINT "ENTER FEET IN DECIMAL FORM"
```

```
630 PRINT
631 PRINT "I.E. ENTER 1/4-FEET AS 0.25"
640 INPUT B1
645 LET B2 = B1 * 0.305
650 LET B3 = B1 * 30.5
655 FOR D = 1 TO 5
656 PRINT
657 NEXT D
660 PRINT B1," FEET =" ;B2," METERS AND ";B3;" CM"
665 FOR F = 1 TO 5
666 PRINT
667 NEXT F
670 PRINT "CONTINUE? PRESS 1 FINISHED? PRESS 2"
680 INPUT B4
690 ON B4 GOTO 600, 10
900 FOR Q = 1 TO 20
905 PRINT
910 NEXT Q
915 PRINT "PROGRAM TO DETERMINE TRANSMISSION LINE DIMENSIONS"'
920 FOR P = 1 TO 4
925 PRINT
930 NEXT P
950 PRINT "WHAT TYPE OF TRANSMISSION LINE?"
955 PRINT "1. FOAM COAXIAL CABLE"
957 PRINT "2. REGULAR COAXIAL CABLE"
959 PRINT "3. TWIN-LEAD CABLE"
965 INPUT B
970 IF B = 1 THEN LET C = 0.8
975 IF B = 2 THEN LET C = 0.66
980 IF B = 3 THEN LET C = 0.82
990 FOR R = 1 TO 5
995 PRINT
1000 NEXT R
1020 PRINT "ENTER FREQUENCY IN KILOHERZ"
1030 INPUT X
1040 PRINT "ENTER WAVELENGTH DESIRED, I.E. 0.125, 0.25, ETC"
1050 INPUT Y
1060 LET L = ((984000 * Y)*C)/X
1065 FOR M = 1 TO 5
1066 PRINT
1067 NEXT M
1070 PRINT Y;"WAVELENGTH IS", %6F2, L;"FEET"
1072 FOR A = 1 TO 5
1073 PRINT
1074 NEXT A
1080 PRINT "CONTINUE? PRESS 1 FINISHED? PRESS 2"
1090 INPUT Z
1100 ON Z GOTO 900, 10
1500 FOR Q = 1 TO 20
1510 PRINT
1515 NEXT Q
1520 PRINT "PROGRAM TO COMPUTE ANTENNA DIMENSIONS"
1530 FOR P = 1 TO 5
1540 PRINT
1550 NEXT P
1560 PRINT "ENTER CORRECT NUMBER FOR ANTENNA DESIRED"
1570 PRINT
1571 PRINT
1573 PRINT "1. FREE-SPACE 1/2-WAVELENGTH"
1576 PRINT "2. FREE-SPACE 1/4-WAVELENGTH"
1578 PRINT "3. 1/2-WAVELENGTH DIPOLE"
1600 INPUT A
1610 ON A GOTO 2500, 3500, 4500
```

```
1620 GOTO 1571
2500 FOR Q = 1 TO 5
2505 PRINT
2510 NEXT Q
2515 PRINT "PROGRAM TO CALCULATE 1/2-WAVELENGTH IN FREE-SPACE"
2520 FOR P = 1 TO 5
2525 PRINT
2530 NEXT P
2535 PRINT "ENTER LOW BAND EDGE IN KILOHERTZ"
2538 PRINT
2540 INPUT F1
2542 PRINT
2544 PRINT "ENTER HIGH BAND EDGE IN KILOHERTZ"
2546 INPUT F2
2548 PRINT
2550 PRINT
2555 PRINT "ENTER NUMBER OF KILOHERTZ INCREMENT"
2560 INPUT F3
2566 LET X = 1
2568 LET L = 492000/F1
2570 PRINT F1;"KHZ";%6F2, L;"FEET"
2575 LET F1 = F1 + F3
2578 IF F1> 52 + F3 THEN GOTO 1520
2580 LET X = X + 1
2585 IF X = 10 THEN GOTO 2595
2590 GOTO 2568
2591 FOR A = 1 TO 5
2592 PRINT
2593 NEXT A
2595 PRINT "CONTINUE? PRESS 1 FINISHED? PRESS 2"
2600 INPUT Y
2605 IF Y = 1 THEN GOTO 2566 ELSE GOTO 1500
3500 FOR Q = 1 TO 20
3505 PRINT
3510 NEXT Q
3515 PRINT "PROGRAM TO COMPUTE 1/4-WAVELENGTH IN FREE-SPACE
3520 FOR P = 1 TO 5
3525 PRINT
3530 NEXT P
3535 PRINT  ENTER LOWER  BAND EDGE, IN KILOHERTZ
3540 INPUT  ?14000
3542 PRINT
3544 PRINT  ENTER HIGH BAND EDGE IN KILOHERTZ
3546 INPUT  ?14400
3548 PRINT
3550 PRINT
3555 PRINT "ENTER NUMBER OF KILOHERTZ INCREMENT"
3560 INPUT F3
3566 LET X = 1
3568 LET L = 246000/F1
3570 PRINT F1;"KHZ";%6F2, L;"FEET"
3575 LET F1 = F1 + F3
3576 IF F1> F2 + F3 THEN GOTO 3595
3579 LET X = X + 1
3585 IF X = 10 THEN GOTO 3595
3590 GOTO 3568
3595 PRINT "CONTINUE? PRESS 1 FINISHED? PRESS 2"
3600 INPUT Y
3605 IF Y = 1 THEN GOTO 3566 ELSE GOTO 1500
4500 FOR Q = 1 TO 20
4505 PRINT
4510 NEXT Q
```

```
4515 PRINT "PROGRAM TO COMPUTE DIMENSIONS OF 1/2-WAVELENGTH DIPOLE"
4520 FOR P = 1 TO 4
4525 PRINT
4530 NEXT P
4535 PRINT "ENTER LOWER BAND EDGE"
4538 PRINT
4540 INPUT F1
4542 PRINT
4544 PRINT "ENTER HIGH BAND EDGE
4546 INPUT F2
4548 PRINT
4550 PRINT
4555 PRINT "ENTER NUMBER OF KILOHERTZ INCREMENT"
4560 INPUT F3
4566 LET X = 1
4568 LET L = 468000/F1
4570 PRINT F1;"KHZ";%6F2, L;"FEET"
4575 LET F1 = F1 + F3
4578 IF F1>F2 + F3 THEN GOTO 4595
4580 LET X = X + 1
4585 IF X = 10 THEN GOTO 4595
4590 GOTO 4568
4595 PRINT "CONTINUE? PRESS 1 FINISHED? PRESS 2"
4600 INPUT Y
4605 IF Y = 1 THEN GOTO 4566 ELSE GOTO 1500
15000 END
```

PROGRAM 8 EXAMPLES

```
EXAMPLES

RUN

WHICH PROGRAM DO YOU WANT?
1. ANTENNA DIMENSIONS
2. FRACTIONAL FEET TO INCHS CONVERSION
3. FEET TO METRIC CONVERSION
4. TRANSMISSION LINE DIMENSIONS
?1


PROGRAM TO COMPUTE ANTENNA DIMENSIONS


ENTER CORRECT NUMBER FOR ANTENNA DESIRED

1. FREE-SPACE 1/2-WAVELENGTH
2. FREE-SPACE 1/4-WAVELENGTH
3. 1/2-WAVELENGTH DIPOLE
?1

PROGRAM TO CALCULATE 1/2-WAVELENGTH IN FREE-SPACE


ENTER LOW BAND EDGE IN KILOHERTZ

?7000
```

```
ENTER HIGH BAND EDGE IN KILOHERTZ
?7300
ENTER NUMBER OF KILOHERTZ INCREMENT
?50
 7000 KHZ        70.29 FEET
 7050 KHZ        69.79 FEET
 7100 KHZ        69.30 FEET
 7150 KHZ        68.81 FEET
 7200 KHZ        68.33 FEET
 7250 KHZ        67.86 FEET
 7300 KHZ        67.40 FEET
 7350 KHZ        66.94 FEET


 ENTER CORRECT NUMBER FOR ANTENNA DESIRED


 1. FREE-SPACE 1/2-WAVELENGTH
 2. FREE-SPACE 1/4-WAVELENGTH
 3. 1/2-WAVELENGTH DIPOLE
 ?2

 PROGRAM TO COMPUTE 1/4-WAVELENGTH IN FREE-SPACE


 ENTER LOWER BAND EDGE IN KILOHERTZ
 ?14000


 ENTER HIGH BAND EDGE IN KILOHERTZ
 ?14400


 ENTER NUMBER OF KILOHERTZ INCREMENT
 ?100
  14000 KHZ        17.57 FEET
  14100 KHZ        17.45 FEET
  14200 KHZ        17.32 FEET
  14300 KHZ        17.20 FEET
  14400 KHZ        17.08 FEET
  14500 KHZ        16.97 FEET
CONTINUE? PRESS 1 FINISHED? PRESS 2


RUN

WHICH PROGRAM DO YOU WANT?
1. ANTENNA DIMENSIONS
2. FRACTIONAL FEET TO INCHS CONVERSION
3. FEET TO METRIC CONVERSION
4. TRANSMISSION LINE DIMENSIONS
?2
PROGRAM TO CONVERT FRACTIONAL FEET TO INCHS


ENTER FRACTIONAL FEET IN DECIMAL FORM

I.E. ENTER 1/4-FEET AS 0.25
?0.335
```

```
.335 FEET =     4.02 INCHS


CONTINUE? PRESS 1 FINISHED? PRESS 2
?2
WHICH PROGRAM DO YOU WANT?
1. ANTENNA DIMENSIONS
2. FRACTIONAL FEET TO INCHS CONVERSION
3. FEET TO METRIC CONVERSION
4. TRANSMISSION LINE DIMENSIONS
?3


PROGRAM TO CONVERT FEET TO METRIC UNITS
ENTER FEET IN DECIMAL FORM

I.E. ENTER 1/4-FEET AS 0.25
?0.357


.357        FEET = .108885      METERS AND  10.8885 CM


CONTINUE? PRESS 1 FINISHED? PRESS 2
?2
WHICH PROGRAM DO YOU WANT?
1. ANTENNA DIMENSIONS
2. FRACTIONAL FEET TO INCHS CONVERSION
3. FEET TO METRIC CONVERSION
4. TRANSMISSION LINE DIMENSIONS
?4


PROGRAM TO DETERMINE TRANSMISSION LINE DIMENSIONS


WHAT TYPE OF TRANSMISSION LINE?
1. FOAM COAXIAL CABLE
2. REGULAR COAXIAL CABLE
3. TWIN-LEAD CABLE
?1

ENTER FREQUENCY IN KILOHERTZ
?27085
ENTER WAVELENGTH DESIRED, I.E. 0.125,0.25,ETC
?.5


.5 WAVELENGTH IS                     14.53 FEET



CONTINUE? PRESS 1 FINISHED? PRESS 2
?2
```

PROGRAM NUMBER 9

This program is for general purpose single-element antennas, such as dipoles and verticals. The constant K is the velocity constant.

```
LIST

 10 LET M = 1
 20 PRINT "INPUT CONSTANT K"
 30 INPUT K
 40 PRINT "INPUT LOWER BAND EDGE IN KILOHERTZ"
 50 INPUT F1
 55 LET G = F1
 60 PRINT "INPUT UPPER BAND EDGE IN KILOHERTZ"
 70 INPUT F2
 80 PRINT "INPUT DESIRED INCREMENTAL FREQUENCY IN KHZ"
 90 INPUT F3
100 PRINT "USE: 1) AMATEUR    2) INT'L BROADCAST"
110 INPUT X
120 PRINT "CONTINUE? PRESS 1"
130 INPUT Y
140 IF Y = 1 THEN GOTO 150
150 FOR Q = 1 TO 5
160 LET F(Q) = F1
170 LET L(Q) = K/F(Q)
180 LET A = L(1)
190 LET B = L(2)
200 LET C = L(3)
210 LET D = L(4)
220 LET E = L(5)
230 LET F1 = F1 + F3
240 NEXT Q
250 IF M > 1 THEN GOTO  310
260 LET M = M + 1
262 PRINT "  ";"BAND:   ";G;" TO ";F2;" KHZ"
270 ON X GOTO 280,300
280 PRINT "  ";"USE: AMATEUR"
282 PRINT
283 PRINT
290 GOTO 310
300 PRINT "  ";"USE: INT'L. BROADCAST"
302 PRINT
303 PRINT
310 PRINT "  ";F(1),"  ";%6F2;A,B,C,D,E
320 IF F1 > F2 THEN GOTO 340
330 GOTO 150
340 END
READY
```

Chapter 13

BASIC Programs for Arccos, Arcsin, and Arctan

Most versions of BASIC used by amateur computerists have only a small number of predefined functions. Usually, the only trigonometric functions offered by these BASICS are Cos X and Sin X. However, many problems that amateur radio operators might want to solve on a computer require arctan, arccos, and arcsin functions. Some of the techniques given in this chapter will allow you to write such programs.

WARNING: PRINCIPLE VALUES

One of the stumbling blocks to writing programs for these functions is the matter of *principle values.* Most of the trigonometric identities and power series that will yield these values are valid only over certain ranges, i.e. the principle values. Sine and cosine functions, for example, have different principle value ranges. The principle values for the sine functions, and the identities using sine, are

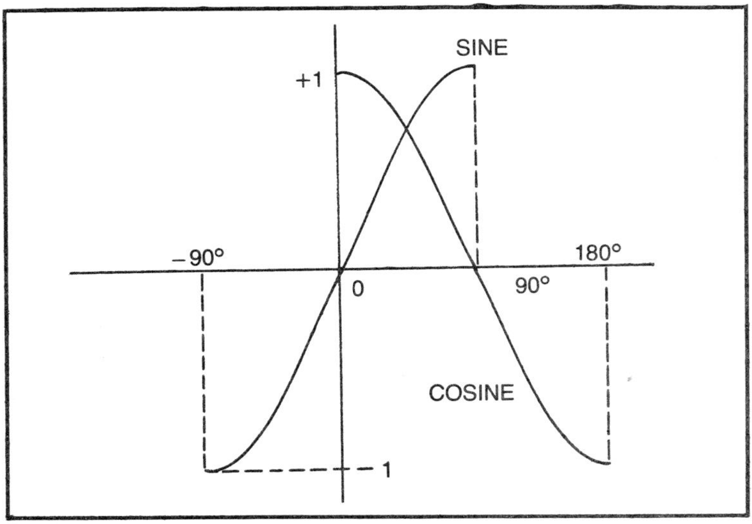

Fig. 13-1. Principle value ranges for sine and cosine functions.

−90 to +90 degrees. The cosine, on the other hand, has a principle value range of 0 to +180 degrees. The only portion of these ranges where there is overlap is 0 to 90 degrees. This is shown in Fig. 13-1. Any routine written to use these functions, must take into account the principle value ranges, or be limited to those values in which the principle values are identical. This means that a simple routine involving sine and cosine must be valid only over the range 0 to 90 degrees.

POWER SERIES

Almost any continuous mathematical function can be represented as a *power series*. The appropriate power series for most of the trigonometric functions are found in most standard books of mathematical tables, and in many algebra/trig and calculus books. The best source, in my opinion, is a book like *CRC Standard Mathematical Tables,* edited by Samual M. Selby (CRC Press). Power series for arccos, arctan and arcsin follow.

- ☐ Arcsin X $= X + \frac{X^3}{2 \cdot 3} + \frac{3X^5}{2 \cdot 4 \cdot 5} + \frac{3 \cdot 5X^7}{2 \cdot 4 \cdot 6 \cdot 7} + \cdots$

- ☐ Arccos X $= \frac{\pi}{2} - \left[\frac{X^3}{2 \cdot 3} + \frac{3X^5}{2 \cdot 4 \cdot 5} + \frac{3 \cdot 5X^7}{2 \cdot 4 \cdot 6 \cdot 7} + \cdots \right]$

- ☐ Arctan X $= -\frac{\pi}{2} - \frac{1}{X} + \frac{1}{3X^3} - \frac{1}{5X^5} + \frac{1}{7X^7} - \cdots$

It is relatively simple to write a BASIC program that will solve one or more of these power series for the arc, if the sine, cosine, or tangent is known. But one *must* provide for the differences in principle value. Examine the following program, which calculates arccos using the arccos power series.

```
1000 DEF FNA(Y)
1010 LET P1 = 3.1415926
1020 IF X = 1 THEN GOTO 2000
1030 IF X = 0 THEN GOTO 2100
1040 IF X = -1 THEN GOTO 2200
1100 LET Y2 = (X+((X'3)/6)+(3*(X'5)/40))
1105 LET Y3 = (15°(X'7)/336)+(105*(X'9)/3024)
1100 LET Y1 = Y2 + Y3
1200 IF X> 0 THEN LET Y = (P ½) - Y1
1300 IF X< 0 THEN LET Y = (P½) + ABS (Y1)
```

```
1500 GOTO 2500
2000 LET Y = 0
2050 GOTO 2500
2100 LET Y = P1/2
2150 GOTO 2500
2200 LET Y = P1
2250 GOTO 2500
2500 RETURN Y
2600 FNEND
READY
```

The first thing we do in this program is set any constants that will be used. In this case, we will need the value π, which is 3.14115926. Next, we test for certain special conditions. We find from examining the power series in the standard math tables book that the series is only valid when X^2 is less than 1, and when the angle is between 0 and 180 degrees. Special conditions exist when X = –1, 0 and +1. These are tested for, and the results set.

If the test conditions are not met, we then calculate the value of the power series. This is not the end of the game, however. If the angle is outside of the range, the angle must be adjusted to within the principle value range.

WHEN YOU HAVE ARCSIN OR ARCTAN FUNCTIONS

If you have an arcsin function in your version of BASIC, then you can use the expression below to calculate arccos:

$$\text{arccos } x = (3.1415926/2) - \text{arcsin } x$$

or

$$\text{arccos } x = (1.5707963) - \text{arcsin } x$$

In the next chapter, we are going to consider a program that calculates the antenna bearing in degrees from your QTH to any point on the face of the earth. In the program is an arccos subroutine that uses the expression:

$$\text{arccos } x = -\text{ATN } (x/\text{SQR } (-x{*}x + 1)) + 1.5708$$

Chapter 14
Calculating Beam Antenna Bearings

The beam antenna, whether quad, Yagi, or delta loop, is said to have gain. But in reality, the gain is nothing more than a simple redirecting of the output energy of the transmitter so that it all radiates in a single direction. The beam antenna will radiate in only a narrow path, usually 15–20 degrees wide. To use a beam antenna, then, we must know in which direction to point the antenna so that the target area of the world receives our signal.

Most of the time, our perception of the proper direction is in error. Consider my own case. I live in the Washington, DC area, which has a latitude approximately the same as southern Spain or Gibraltar. By conventional wisdom, one would think that to hit a station in southern Spain, on approximately the same latitude as my station, that I would have to aim the antenna due east or 90 degrees from true north. But a signal sent due east from my location would reach its first landfall not in Spain—or in fact anywhere in Europe—but in Angola on the west coast of Africa!

The problem is that radio signals travel the shortest path between two points. On a plane surface, this is a straight line, but on a sphere it is an arc called a *great circle* path. We may view the earth as a sphere for purposes of making beam heading calculations, and then apply the ordinary rules for spherical trigonometry. Before you turn off, saying that math scares you, however, listen awhile and let us teach you to make these calculations. The arithmetic is not all that complicated, and we are going to give you a program in BASIC to calculate the bearings between your QTH and any point on the surface of the earth.

Incidentally, if you think that you can solve the problem of finding beam headings from some other source, then let me caution you that doing it yourself may still be better. First, the great circle charts in the amateur publications are usually valid only at two or three points in the USA. They assume that everyone lives within 100 miles of Washington, DC, Wichita, KS, or San Francisco, CA. Everyone else must interpolate and hope for the best. You might also say "but I can send a fee to one of those computer outfits that advertise in the amateur magazines and get a printout customized to my location!" True, but by doing it my way, you can calculate the bearings to the cities or countries that interest *you*, not those that some computer jockey thinks you should be interested in. Another point is that the fee they charge will pay for about half of this book. So this one chapter pays for most of the rest of the book, and gives you the ability to program your own microcomputer to find antenna bearings.

LATITUDE AND LONGITUDE

Consider Fig. 14-1. We show here two locations on the surface of the spherical earth. Location A is your location, while B is the location of the other station—country or city. We can compute the bearings from true north required to aim the antenna at point A in order to send a signal to point B by knowing the latitude and longitude for each station.

Latitudes are measured from the equator, which is assigned a number 0 degrees. The North pole is given the number +90 degrees, and the south pole is given the number −90 degrees. Note that all north latitudes are taken to be positive (+) and all south

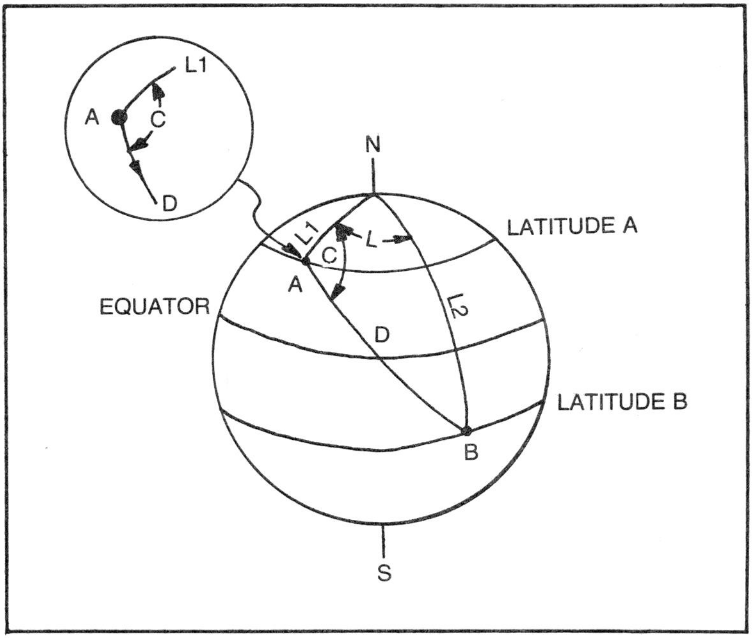

Fig. 14-1. Earth's surface.How do we find the bearing from true north (angle C) from point A to point B?

latitudes are taken to be negative (−). The latitudes in all cases run from 0 to 90 or 0 to −90 degrees.

Longitude measures locations east and west of a certain line drawn through the north pole, south pole and the observatory in Greenwich, England. This line is called the prime meridian and is assigned to the zero degrees longitude. Points east and west of the prime meridian are measured in degrees. West longitudes are taken to be positive (+) and east longitudes are considered negative (−).

We can precisely locate any point on the surface of the earth if we know its exact latitude and longitude. Furthermore, we may calculate the distance between two points if we know the latitude and longitudes of both points. The bearings from true north from one of the points to the other is also determined from this same information.

In Fig. 14-1 we have used the same definitions as other articles on bearings calculations:

A = your latitude in degrees
B = the latitude of the other station, city, or country in degrees
L1 = your longitude in degrees
L2 = the longitude of the other station, city, or country in degrees
D = the arc length along the short path in degrees
C = the heading from true north between points A and B
L = the difference L1 − L2. If L is greater than +180 degrees, or less than −180 degrees, then L must be corrected to fall within that range. If L is greater than +180 degrees, then *subract* 360, but if L is less than −180 degrees, then *add* 360 degrees. In other words, do whatever is needed to bring L into the range ± 180 degrees.

The bearing, C, might also require correction. If the final, corrected value of L is positive, then C = C. But if the final, corrected value of L is negative, then C = 360 − C.

Two equations are needed to find the bearings:

$$\text{Cos } D = (\text{Sin } A)\,(\text{Sin } B) + (\text{Cos } A)\,(\text{Cos } B)\,(\text{Cos } L)$$

$$\text{Cos } C = \frac{(\text{Sin } B) - (\text{Sin } A)\,(\text{Cos } D)}{(\text{Cos } A)\,(\text{Sin } D)}$$

□ **Example**

Calculate the antenna bearings from Washington, D.C. to Rio de Janeiro, Brazil.

Washington, D.C.: Lat. A = 38.9 degrees Long. L1 = 77 degrees
Rio de Janiero: Lat. B = −22.8 degrees Long. L2 = 43.2 degrees

$$L = L1 - L2 = 77 - 43.2 = +33.8$$

Note that L is less than +180 and more than −180, so no correction is needed. Also, since L is positive, the final value calculated for C below will not require any correction.

$\text{Cos D} = \text{Sin}(38.9)\,\text{Sin}(-22.9) + \text{Cos}(38.9)\,\text{Cos}(-22.9)\,\text{Cos}(33.8)$

$\text{Cos D} = (0.62796)\,(-0.38912) + (0.7782)\,(0.9212)\,(0.83096)$

$\text{Cos D} = 0.35138$

$\text{D} = \text{Arccos } 0.35138 = 69.4 \text{ degrees}$

$$\text{Cos C} = \frac{\text{Sin}(-22.9) - \text{Sin}(38.9)\,\text{Cos}(69.4)}{\text{Cos}(38.9)\,\text{Sin}(69.4)}$$

$$\text{Cos C} = \frac{(-0.38912) - (0.62115)\,(0.3518)}{(0.7782)\,(0.9361)}$$

$$\text{Cos C} = \frac{-0.6076}{0.7285}$$

$\text{Cos C} = -0.8341$

$\text{C} = \text{Arccos } 0.8341 = 147 \text{ degrees}$

If we want to know the distance to the other station, then we use the arc length D. Each degree of arc length along the path is equal to 60 nautical miles. To find nautical miles, we merely multiply arc length D by the constant 60. But if we want statute miles (the kind we travel on land), then multiply nautical miles by 1.15078, and if kilometer measure is desired multiply by 1.852. In the example above, arc length D was found to be 69.4 degrees, so D = 69.4 degrees.

Distance is:

$69.4 \times 60 = 4164$ nautical miles
$69.4 \times 60 \times 1.15078 = 4818$ statute miles
$69.4 \times 60 \times 1.852 = 7712$ kilometers

BASIC PROGRAM

Following is a program in BASIC that will perform the calculations given above if you supply the latitudes and longitudes of both stations. This program was written in the version of BASIC used in the Apple II microcomputer. Note that lines 640 and 750 use the ATN (arctan) function, which is not universally available in all versions of BASIC used in microcomputers. It is available in Apple

BASIC, but in the Digital Group BASIC, of which I am also familiar, it is not available. Consult Chapter 13 for appropriate algorithms for calculating an arccos or arctan in those BASICs that lack this function. You can program it as a subroutine, or as a "user defined function," such as FNA.

Note that this program converts the latitudes and longitudes into radian form. This is done because most computer functions to calculate trig values use radian notation. The units are converted back to degrees prior to printout.

You may wish to redesign portions of this program to better suit your own needs. For example, you may want to preprogram the values of A and L1, your own latitude and longitude, because it is a reasonable assumption that your house won't move around much.

You may also wish to redesign the portion of the program that inputs the latitude and longitude of the other station. One way is to dimension variables B and L2 as an array. This array can then be placed in a FOR, NEXT loop to allow you to input the coordinates and names of a large number of cities. This would allow you to calculate the bearings to all major cities and countries(and some not so major . . . DXers do not usually care how large a "country" is, and in fact, some "countries" are merely islands).

```
LIST

 50  REM  SET CONSTANTS
100  LET R = 57.295779
110  LET P1 = 3.1415926
115  FOR Q = 1 TO 5
116  PRINT
117  NEXT Q
200  REM  INPUT YOUR LATITUDE AND LONGITUDE
210  PRINT "ENTER YOUR LATITUDE IN DECIMAL DEGREES"
220  PRINT "I.E. 38.9 DEGREES"
230  INPUT A
235  FOR Q = 1 TO 5
240  PRINT
245  NEXT Q
250  PRINT "ENTER YOUR LONGITUDE IN DECIMAL DEGREES"
260  INPUT L1
270  FOR Q = 1 TO 5
272  PRINT
275  NEXT Q
300  REM  INPUT LATITUDE AND LONGITUDE OF310 REM OTHER CITY OR COUNTRY
310  FOR Q = 1 TO 5
312  PRINT
315  NEXT Q
320  PRINT "ENTER NAME OF COUNTRY OR CITY"
330  INPUT B$
335  FOR Q = 1 TO 5
336  PRINT
337  NEXT Q
340  PRINT "ENTER LATITUDE OF ";B$
```

```
350 INPUT B
355 FOR Q = 1 TO 5
356 PRINT
357 NEXT Q
360 PRINT "ENTER LONGITUDE OF ";B$
370 INPUT L2
400 REM  CALCULATE LONGITUDE DIFFERENCE
410 LET L3 = L1 - L2
420 IF  ABS (L3) = 180 THEN  GOTO 500
430 IF  ABS (L3) < 180 THEN  GOTO 500
440 IF L3 > 180 THEN  LET L3 = L3 - 360
450 IF L3 <  - 180 THEN  LET L3 = L3 + 360
490 REM  CONVERT TO RADIANS
500 LET L = L3 / R
510 LET A = A / R
520 LET B = B / R
600 REM  CALCULATE LONGITUDE DIFFERENCE
610 LET D1 =  SIN (A) *  SIN (B)
620 LET D2 =  COS (A) *  COS (B) *  COS (L)
622 REM
623 REM
625 REM  D3 IS COS D
626 REM
630 LET D3 = D1 + D2
640 LET D =  -  ATN (D3 /  SQR ( - D3 * D3 + 1)) + 1.5708
700 REM  CALCULATE BEARING
710 LET C1 =  SIN (B)
720 LET C2 =  SIN (A) * D3
730 LET C3 =  COS (A) *  SIN (D)
740 LET C4 = ((C1 - C2) / C3)
745 REM
746 REM
747 REM  CALCULATE BEARING
748 REM
749 REM
750 LET C =  -  ATN (C4 /  SQR ( - C4 * C4 + 1)) + 1.5708
755 REM
756 REM
760 LET I = L3
770 IF I > 0 THEN  LET C = C
780 IF I < 0 THEN  LET C = (2 * P1) - C
800 REM  RECONVERT ALL ANGLES TO DEGREES
810 LET C = C * R
820 LET A = A * R
830 LET B = B * R
840 LET D = D * R
900 REM  ROUND OFF BEARING
910 LET M = C -  INT (C)
920 IF M > 0.5 THEN  LET C =  INT (C) + 1
930 IF M = 0 THEN  LET C = C
940 IF M < 0.5 THEN  LET C =  INT (C)
1000 REM  CALCULATE DISTANCE
1005 REM  S1 IS MILES, S2 IS KILOMETERS
1010 LET S1 = D * 60 * 1.15078
1020 LET S2 = D * 60 * 1.852
1100 REM  ROUND OFF DISTANCES
1110 LET G = S1 -  INT (S1)
1115 IF G = 0 THEN  LET S1 = S1
1120 IF G > 0.5 THEN  LET S1 =  INT (S1) + 1
1125 IF G < 0.5 THEN  LET S1 =  INT (S1)
```

```
1130  LET H = S2 -  INT (S2)
1135  IF H = 0 THEN  LET S2 = S2
1140  IF H > 0.5 THEN  LET S2 =  INT (S2) + 1
1150  IF H < 0.5 THEN  LET S2 =  INT (S2)
2000  REM  PRINT OUTPUT STUFF
2005  PRINT
2006  PRINT
2020  PRINT B$
2022  PRINT "LAT. ";B;"  ";"LONG. ";L2
2030  PRINT S1;" MILES",S2;"KM"
2040  PRINT "BEARING: ";C;" DEGREES"

]RUN

ENTER YOUR LATITUDE IN DECIMAL DEGREES
I.E. 38.9 DEGREES
?38.9

ENTER YOUR LONGITUDE IN DECIMAL DEGREES
?77

ENTER NAME OF COUNTRY OR CITY
?SYDNEY AUSTRALIA

ENTER LATITUDE OF SYDNEY AUSTRALIA
?-33.9

ENTER LONGITUDE OF SYDNEY AUSTRALIA
?-151.2

SYDNEY AUSTRALIA
LAT. -33.9  LONG. -151.2
9759 MILES      15705KM
BEARING: 262 DEGREES
```

Chapter 15
Antenna Transmission Lines & SWR

It is rare that one can locate the rf power amplifier of their transmitter right at the feedpoint of the antenna. Although some VHFers seeking maximum output power for certain exotic communications will do this, it is not exactly practical for most amateurs. For most of us, the rf power amplifier will be either inside the transceiver housing, or right along side of the transceiver on the operating table. A *transmission line* is needed to carry the rf power generated by the transmitter to the antenna, and to bring signals received by the antenna back to the receiver.

In the case of certain crude antennas such as the long wire, or the windom (a form of long wire), the "transmission line" is a piece of No. 12 or No. 14 wire, usually of the same sort as used in the antenna itself. But more complex antennas require better transmission lines.

The transmission line for a balanced antenna such as a dipole might well be merely a pair of conductors placed side-by-side as in Fig. 15-1. The transmission line differs from merely a pair of wires, however, in that it has certain properties. We will consider the velocity factor and the surge impedance of the line.

The velocity factor of the transmission line is derived from the fact that radio waves will propogate at the speed of light (300,000,000 meters per second) in free-space, but will have only a fraction of that velocity when in the transmission line. This fraction, of course, is less than unity (1), and will be expressed as a decimal.

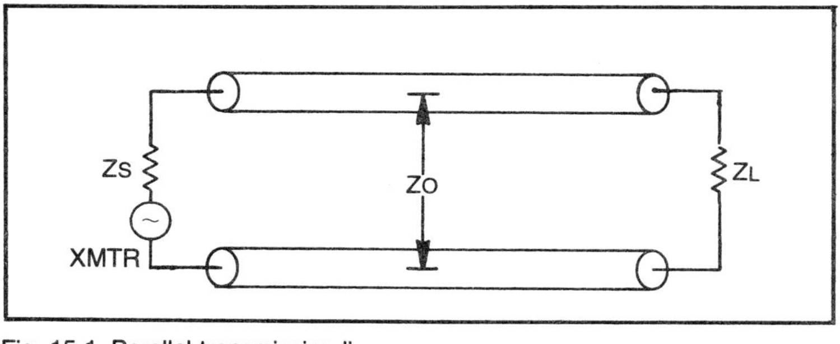

Fig. 15-1. Parallel transmission line.

Table 15-1 shows the velocity factors of several popular types of transmission line.

Cable	Velocity Factor
Regular coax	0.66
Foam-filled coax	0.80
300 – ohm twinlead	0.81-0.82
Open feeders (i.e. ladder line)	0.80-0.83

Table 15-1. Velocity factors of transmission lines.

The velocity factor tells us the fraction of the speed of light that the radio signal will travel through the cable. For example, in regular polyethylene dielectric coaxial cable, in which velocity factor V is 0.66, the speed of the signal is 0.66 times the speed of light, or

$V = (0.66)\ (3 \times 10^8\ \text{m/s})$
$V = 1.98 \times 10^8\ \text{m/s}$
$V = 198{,}000{,}000$ meters/second

The equations that we used in determining wavelengths of antennas must be modified for finding an *electrical* length of transmission line. Note that the physical length of a piece of transmission line is longer than the electrical length. If you measure the physical length of a piece of polyethylene coaxial cable that is said to be an electrical half wavelength at 27 MHz, you would not find the length to be the same as calculated from the 492/F expression. The electrical length of transmission line is found from the expressions:

Half wavelength:

$$L = (492 \times V)\ /F \qquad V = \frac{LF}{492}$$

Quarter wavelength:

$$L = (246 \times V)\ /F \qquad V = \frac{LF}{246}$$

where L is the length in feet, F is the frequency in megahertz (MHz) and V is the velocity factor of the transmission line.

☐ **Example**

Find the physical length of a piece of polythylene dielectric coaxial cable that is to be an electrical half wavelength at 14.25 MHz.

☐ **Solution**

$$L = (492 \times V)/F$$
$$L = (492 \times 0.66)/14.25$$
$$L = 325/14.25 = 22.8 \text{ feet}$$

A half wavelength from the expression without the velocity factor term would yield 34.5 feet, or about the normal length of a 20-meter dipole.

The *surge impedance,* also called *characteristic impedance*, of a transmission line is associated with the distributed inductance and capacitance of the line per unit of length:

$$Z_0 = \sqrt{L_0/C_0}$$

where Zo is the surge impedance in ohms, Lo is the inductance per unit length, and Co is the capacitance per unit length.

The normal values for surge impedance are different for the different types of transmission line. For coaxial cable, for example, the values of Zo run from 10 ohms to 150 ohms. Two popular impedances are 52 ohms (RG-58 and RG-8), and 73 ohms (RG-59 and RG-11). Parallel conductor transmission lines can have impedances in the range 150 ohms to 1000 ohms. Values of 150, 300 and 600 ohms are common.

The impedance shown as Z_L in Fig. 15-1 is the load impedance, and represents the impedance of the antenna feedpoint. This impedance might be anything from 5 ohms to 1000 ohms, so obviously the correct type of line must be used, or a matching device be provided.

Antenna impedance is of the form:

$$Z_L = R \pm jX$$

where Z_L is the antenna feedpoint impedance, R is the resistive component, j is the operator (−1) ½ and X is the reactive component, either X_L or Xc.

In a perfectly resonant antenna system, the inductive and capacitive reactances are equal, so the reactance term X is canceled out, leaving only the resistance. This value, R, is the radiation resistance of the antenna and is the quotient of the feedpoint voltage divided by the feedpoint current. Most practical antennas, however, do not have a zero reactive component, but will have some value between 5 and 100 ohms. The lower the value, however, the better.

The matter of *standing waves,* and the standing waves ratio, or SWR, is very important to the proper functioning of the antenna system and the safety of the transmitter (high SWR can and does blow out rf power amplifiers). But SWR has caused a lot of misapprehension among amateurs and CBers alike. Let us try to clear up a little bit of the mess by considering standing waves. We will first use a mechanical model that graphically shows the property called standing waves.

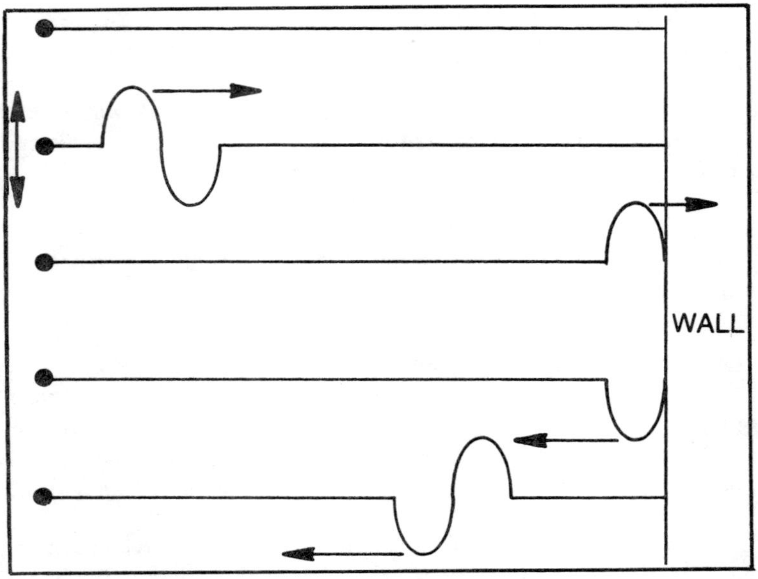

Fig. 15-2. Rope anology for standing waves.

Our first anology is similar to what happens in a transmission line when a single pulse is applied. Consider Fig. 15-2. If a rope were attached at one end to an immovable wall, and is pulled taut, we would be able to impart a wave to the rope by a quick flick of the wrist. The wave would travel *forward* along the rope until it meets the wall. When it hits the wall, the wave will be *reflected* back towards the "transmitter" end. Note that a reversal of phase took place on reflection. In a perfect rope, the wave would continue to oscillate back and forth indefinitely, but in real ropes losses cause the oscillation to diminish quickly.

Now let us consider what happens when a continuous signal is applied to the rope. See Fig. 15-3. If the transmitter end of the rope is moved up and down, it will impart a sine wave to the rope. Immediately after reflection, there will be two waves on the rope—the forward wave and the reflected wave. The wave at any given point will be the sum of these two (since the rope cannot be in two places at once). If the rope is moved in exactly the correct frequency, we will see a phenomenon called standing waves in which the rope appears to stand still, forming the characteristic nodes and antinodes of Fig. 15-3.

In antenna systems we see this same phenomenon, but with electrical waves rather than mechanical waves. Figure 15-4 graphs the current or voltage along the line as a function of distance *back* from the load end, or the antenna feedpoint. Interestingly enough, the behavior of the wave is almost the same for an open antenna as for a shorted antenna. Only the physical location of the nodes and antinodes changes. The solid line in Fig. 15-4A shows the situation when the transmission line is shorted at the load end ($Z_L = 0$). In this case, the voltage applied to the antenna feedpoint is zero, and we find other points of zero voltage at half-wavelength intervals back down the line. Similar behavior is noted when the transmission line sees an open load. In this case, however, the voltage at the feedpoint of the antenna (or the point where it should be) is maximum, so an antinode is reflected at half-wavelength intervals back down the line.

The situation in a perfect antenna system is shown in Fig. 15-4B. Here the load impedance of the antenna is matched exactly to the surge impedance of the transmission line. The line is said to be "flat" for obvious reasons under this circumstance.

Very few real antennas are either as bad as Fig. 15-4A or as good as Fig. 15-4B. Most of them present a situation more like Fig. 15-4C, in which $Z_L \neq Z_o$, but neither impedance is zero. In this case, some of the power is reflected back down the line toward the transmitter end and some is radiated. Of course, the less power that is reflected, the more we can radiate to produce a stronger signal. This is supposedly the reason why so much attention is paid to SWR in transmitter systems, especially on CB where most operators run less than 5 watts.

The *standing wave ratio* is a measure of how well the antenna and transmission line are matched. We actually have several different methods for measuring SWR based on the formulas below:

1) $SWR = Emax/Emin$
2) $SWR = Imax / Imin$
3) $SWR = Z_L/Z_o$ if Z_o is less than Z_L
4) $SWR = Z_o/Z_L$ (if Z_o is greater than Z_L)
5) $SWR = (Ef + Er)/(Ef - Er)$
6) $SWR = (1 + \sqrt{Pr/Pf})/(1 - \sqrt{Pr/Pf})$

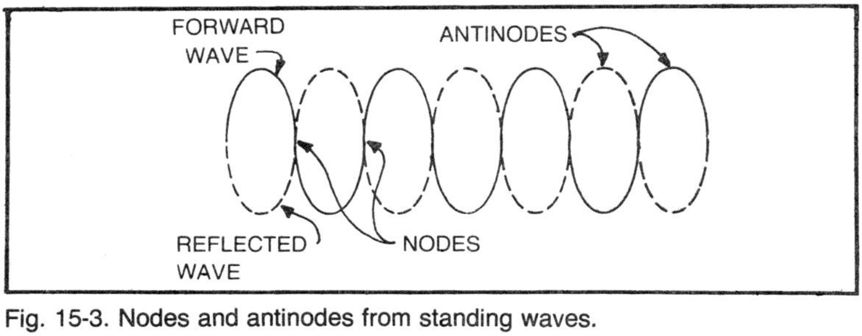

Fig. 15-3. Nodes and antinodes from standing waves.

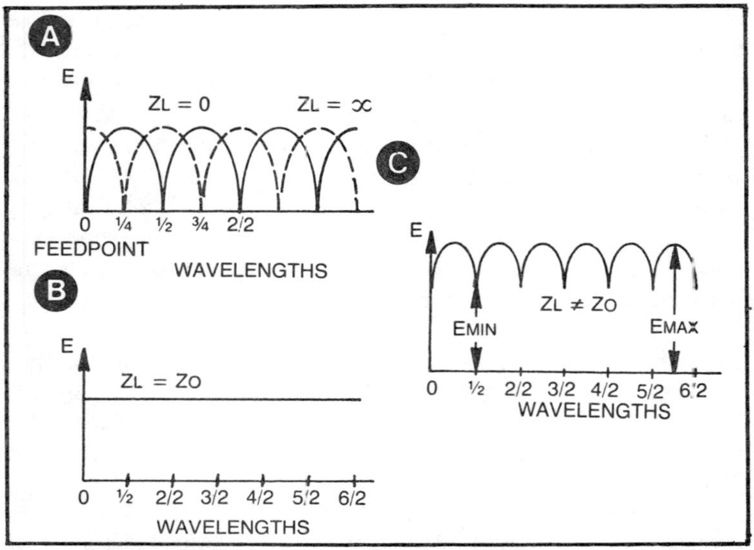

Fig. 15-4. Standing wave situation along a transmission line from the antenna end under various situations. At (A), $Z_L = 0$ and $Z_L = \infty$, at (B) $Z_L = Z_o$, and at (C) $Z_L \neq Z_o$.

where Zo is the surge impedance of the transmissional line, Z_L is the antenna impedance, E_{max} is the maximum voltage along the transmission line, E_{min} is the minimum voltage along the transmission line, I_{max} is the maximum current along the transmission line, I_{min} is the minimum current along the transmission line, E_f is the forward component of the voltage at any given point along the transmission line, E_r is the reflected voltage component at any given point along the transmission line, P_f is the forward power in the line, P_r is the reflected power in the line.

SWR measurement instruments will be discussed in Chapter 16, but right now I want to make a couple of things known about the technique of measurement and proper adjustment of the system for best SWR, which is 1:1. If you are using a true rf wattmeter, and plug in equation 6, then the accuracy of the measurement is almost unaffected by the length of the antenna transmission line. But if you use one of the meters based on current or voltage in the transmission line, such as the often used equation 5, then the transmission line length is *very* important. Most amateur and not a few professional SWR meters are based on samplings of the forward and reflected voltage or current at a point in the line. In this case, the only points at which the instrument calibration is valid is the antenna feedpoint or at electrical half wavelengths back down the line from the feedpoint. This was shown graphically in Fig. 15-4C. If the antenna line length is unknown, or if it is known to be something other than an electrical half wavelength, then don't quote your SWR reading to me! I won't believe you. The readings have meaning *only* when the transmission line is of a length equal to an integer (i.e. 1, 2, 3...) multiple of a half wavelength. Don't forget, we find the electrical half wavelength of the transmission line from the expression $L = (492 \times V)/F$.

If you do not know the length of the transmission lne, then you can still use the SWR meter to adjust the antenna system. The trick is to get the minimum SWR obtainable, regardless of what the numbers might be. But do not use the old CBers trick of cutting the transmission line for minimum SWR. This is a fool's waste of time. The SWR is adjusted *only* by matching the antenna—usually by adjusting length—to the transmission line. Cutting the line to obtain minimum SWR is only apparently true, and is due to the fact that most SWR meters are voltage or current operated. A power-operated meter would not show that phenomenon.

Never trim the length of the coax to obtain minimum SWR! You are not accomplishing anything. Instead, adjust the antenna length, or tuning controls, to obtain a good match between the antenna and the coax.

TABLE 15-2
HALF AND QUARTER WAVELENGTHS
3500 - 4000 KILOHERTZ
USE: AMATEUR

			VEL.FAC.		
FORMAT: FREQ.	1/4-W	1/2-W	0.66,	0.80,	0.82
3500	66.8	133.7	88.2	106.9	109.6
3520	66.4	132.9	87.7	106.3	109
3540	66.1	132.2	87.2	105.7	108.4
3560	65.7	131.4	86.7	105.1	107.7
3580	65.3	130.7	86.2	104.5	107.1
3600	65	130	85.7	104	106.5
3620	64.6	129.2	85.3	103.4	106
3640	64.2	128.5	84.8	102.8	105.4
3660	63.9	127.8	84.3	102.2	104.8
3680	63.5	127.1	83.9	101.7	104.2
3700	63.2	126.4	83.4	101.1	103.7
3720	62.9	125.8	83	100.6	103.1
3740	62.5	125.1	82.5	100.1	102.6
3760	62.2	124.4	82.1	99.5	102
3780	61.9	123.8	81.7	99	101.5
3800	61.5	123.1	81.2	98.5	100.9
3820	61.2	122.5	80.8	98	100.4
3840	60.9	121.8	80.4	97.5	99.9
3860	60.6	121.2	80	96.9	99.4
3880	60.3	120.6	79.6	96.4	98.9
3900	60	120	79.2	96	98.3
3920	59.6	119.3	78.7	95.5	97.8
3940	59.3	118.7	78.3	95	97.4
3960	59	118.1	78	94.5	96.9
3980	58.7	117.5	77.6	94	96.4
4000	58.5	117	77.2	93.6	95.9

TABLE 15-3
HALF AND QUARTER WAVELENGTHS
7000 - 7300 KILOHERTZ
USE: AMATEUR

				VEL.FAC.	
FORMAT: FREQ.	1/4-W	1/2-W	0.66,	0.80,	0.82
7000	33.4	66.8	44.1	53.4	54.8
7020	33.3	66.6	43.9	53.3	54.6
7040	33.2	66.4	43.8	53.1	54.5
7060	33.1	66.2	43.7	53	54.3
7080	33	66.1	43.6	52.8	54.2
7100	32.9	65.9	43.5	52.7	54
7120	32.8	65.7	43.3	52.5	53.8
7140	32.7	65.5	43.2	52.4	53.7
7160	32.6	65.3	43.1	52.2	53.5
7180	32.5	65.1	43	52.1	53.4
7200	32.5	65	42.8	52	53.2
7220	32.4	64.8	42.7	51.8	53.1
7240	32.3	64.6	42.6	51.7	53
7260	32.2	64.4	42.5	51.5	52.8
7280	32.1	64.2	42.4	51.4	52.7
7300	32	64.1	42.3	51.2	52.5

TABLE 15-4
HALF AND QUARTER WAVELENGTHS
14000 - 14450 KILOHERTZ
USE: AMATEUR

			VEL.FAC.		
FORMAT: FREQ.	1/4-W	1/2-W	0.66,	0.80,	0.82
14000	16.7	33.4	22	26.7	27.4
14020	16.6	33.3	22	26.7	27.3
14040	16.6	33.3	21.9	26.6	27.3
14060	16.6	33.2	21.9	26.6	27.2
14080	16.6	33.2	21.9	26.5	27.2
14100	16.5	33.1	21.9	26.5	27.2
14120	16.5	33.1	21.8	26.5	27.1
14140	16.5	33	21.8	26.4	27.1
14160	16.5	33	21.8	26.4	27.1
14180	16.5	33	21.7	26.4	27
14200	16.4	32.9	21.7	26.3	27
14220	16.4	32.9	21.7	26.3	26.9
14240	16.4	32.8	21.6	26.2	26.9
14260	16.4	32.8	21.6	26.2	26.9
14280	16.3	32.7	21.6	26.2	26.8
14300	16.3	32.7	21.6	26.1	26.8
14320	16.3	32.6	21.5	26.1	26.7
14340	16.3	32.6	21.5	26.1	26.7
14360	16.2	32.5	21.5	26	26.7
14380	16.2	32.5	21.4	26	26.6
14400	16.2	32.5	21.4	26	26.6
14420	16.2	32.4	21.4	25.9	26.6
14440	16.2	32.4	21.3	25.9	26.5

TABLE 15-5
HALF AND QUARTER WAVELENGTHS
21000 - 21450 KILOHERTZ
USE: AMATEUR

			VEL.FAC.		
FORMAT:FREQ.	1/4-W	1/2-W	0.66,	0.80,	0.82
21000	11.1	22.2	14.7	17.8	18.2
21020	11.1	22.2	14.6	17.8	18.2
21040	11.1	22.2	14.6	17.7	18.2
21060	11.1	22.2	14.6	17.7	18.2
21080	11.1	22.2	14.6	17.7	18.2
21100	11	22.1	14.6	17.7	18.1
21120	11	22.1	14.6	17.7	18.1
21140	11	22.1	14.6	17.7	18.1
21160	11	22.1	14.5	17.6	18.1
21180	11	22	14.5	17.6	18.1
21200	11	22	14.5	17.6	18.1
21220	11	22	14.5	17.6	18
21240	11	22	14.5	17.6	18
21260	11	22	14.5	17.6	18
21280	10.9	21.9	14.5	17.5	18
21300	10.9	21.9	14.5	17.5	18
21320	10.9	21.9	14.4	17.5	18
21340	10.9	21.9	14.4	17.5	17.9
21360	10.9	21.9	14.4	17.5	17.9
21380	10.9	21.8	14.4	17.5	17.9
21400	10.9	21.8	14.4	17.4	17.9
21420	10.9	21.8	14.4	17.4	17.9
21440	10.9	21.8	14.4	17.4	17.8

?1

TABLE 15-6
HALF AND QUARTER WAVELENGTHS
28500 - 29000 KILOHERTZ
USE: AMATEUR

FORMAT: FREQ.	1/4-W	1/2-W	0.66,	0.80,	VEL.FAC. 0.82
28500	8.2	16.4	10.8	13.1	13.4
28520	8.2	16.4	10.8	13.1	13.4
28540	8.1	16.3	10.8	13.1	13.4
28560	8.1	16.3	10.8	13.1	13.4
28580	8.1	16.3	10.8	13.1	13.4
28600	8.1	16.3	10.8	13	13.4
28620	8.1	16.3	10.7	13	13.4
28640	8.1	16.3	10.7	13	13.3
28660	8.1	16.3	10.7	13	13.3
28680	8.1	16.3	10.7	13	13.3
28700	8.1	16.3	10.7	13	13.3
28720	8.1	16.2	10.7	13	13.3
28740	8.1	16.2	10.7	13	13.3
28760	8.1	16.2	10.7	13	13.3
28780	8.1	16.2	10.7	13	13.3
28800	8.1	16.2	10.7	13	13.3
28820	8.1	16.2	10.7	12.9	13.3
28840	8.1	16.2	10.7	12.9	13.3
28860	8.1	16.2	10.7	12.9	13.2
28880	8.1	16.2	10.6	12.9	13.2
28900	8	16.1	10.6	12.9	13.2
28920	8	16.1	10.6	12.9	13.2
28940	8	16.1	10.6	12.9	13.2
28960	8	16.1	10.6	12.9	13.2
28980	8	16.1	10.6	12.9	13.2
29000	8	16.1	10.6	12.9	13.2

TABLE 15-7
HALF AND QUARTER WAVELENGTHS
29000 - 29500 KILOHERTZ
USE: AMATEUR

FORMAT: FREQ.	1/4-W	1/2-W	0.66,	0.80,	VEL.FAC. 0.82
29000	8	16.1	10.6	12.9	13.2
29020	8	16.1	10.6	12.9	13.2
29040	8	16.1	10.6	12.8	13.2
29060	8	16.1	10.6	12.8	13.2
29080	8	16	10.6	12.8	13.1
29100	8	16	10.6	12.8	13.1
29120	8	16	10.6	12.8	13.1
29140	8	16	10.5	12.8	13.1
29160	8	16	10.5	12.8	13.1
29180	8	16	10.5	12.8	13.1
29200	8	16	10.5	12.8	13.1
29220	8	16	10.5	12.8	13.1
29240	8	16	10.5	12.8	13.1
29260	7.9	15.9	10.5	12.7	13.1
29280	7.9	15.9	10.5	12.7	13.1
29300	7.9	15.9	10.5	12.7	13
29320	7.9	15.9	10.5	12.7	13
29340	7.9	15.9	10.5	12.7	13
29360	7.9	15.9	10.5	12.7	13
29380	7.9	15.9	10.5	12.7	13
29400	7.9	15.9	10.5	12.7	13
29420	7.9	15.9	10.4	12.7	13
29440	7.9	15.8	10.4	12.7	13
29460	7.9	15.8	10.4	12.7	13
29480	7.9	15.8	10.4	12.7	13
29500	7.9	15.8	10.4	12.6	13

?1

TABLE 15-8
HALF AND QUARTER WAVELENGTHS
29500 - 30000 KILOHERTZ
USE: AMATEUR

FORMAT: FREQ.	1/4-W	1/2-W	VEL.FAC. 0.66,	0.80,	0.82
29500	7.9	15.8	10.4	12.6	13
29520	7.9	15.8	10.4	12.6	13
29540	7.9	15.8	10.4	12.6	12.9
29560	7.9	15.8	10.4	12.6	12.9
29580	7.9	15.8	10.4	12.6	12.9
29600	7.9	15.8	10.4	12.6	12.9
29620	7.9	15.8	10.4	12.6	12.9
29640	7.8	15.7	10.4	12.6	12.9
29660	7.8	15.7	10.4	12.6	12.9
29680	7.8	15.7	10.4	12.6	12.9
29700	7.8	15.7	10.4	12.6	12.9
29720	7.8	15.7	10.3	12.5	12.9
29740	7.8	15.7	10.3	12.5	12.9
29760	7.8	15.7	10.3	12.5	12.8
29780	7.8	15.7	10.3	12.5	12.8
29800	7.8	15.7	10.3	12.5	12.8
29820	7.8	15.6	10.3	12.5	12.8
29840	7.8	15.6	10.3	12.5	12.8
29860	7.8	15.6	10.3	12.5	12.8
29880	7.8	15.6	10.3	12.5	12.8
29900	7.8	15.6	10.3	12.5	12.8
29920	7.8	15.6	10.3	12.5	12.8
29940	7.8	15.6	10.3	12.5	12.8
29960	7.8	15.6	10.3	12.4	12.8
29980	7.8	15.6	10.3	12.4	12.8
30000	7.8	15.6	10.2	12.4	12.7

Chapter 16
Simple Antenna Instrumentation

Every serious radio communications hobbiest should own at least some form of antenna test equipment. This does not have to mean a large investment, and indeed much of the instrumentation can be homebrewed. Many radio amateurs use homebrew instruments constructed of junk parts, and these produce as good results as very expensive commercial equipment.

Antenna engineers will use some high-priced and exotic equipment when making antenna measurements. But their needs for equipment are based on their needs for accurate, numerical data concerning the antenna performance. In amateur and CB work, we can use more crude types of equipment because our goal is not to understand the physics of the beast, but to just make it work well.

We will be using several different types of antenna instruments when testing or troubleshooting amateur antenna systems. In some cases, the instrument will be a *thermocouple rf ammeter*, while in others it will be a *true rf wattmeter*. In most cases, the SWR of an antenna is measured by a simple voltage-or current-operated *bridge* circuit that compares reflected and forward components at a given point (wherein lies the problem of transmission line length). In still other cases, we will want to use an *impedance bridge* that actually measures the resistive portion of the antenna impedance, or a *noise bridge*, which is a fantastically useful instrument—and low cost, too.

THERMOCOUPLE RF AMMETERS

The measurement of rf current is almost impossible using most rectifier-type AC meters. In order to measure the true rms value of the current flowing in a transmission line, it is necessary to use a heat-operated device. The *thermocouple* is a device that produces a voltage when heated, and the voltage is proportional to the temperature of the thermocouple junction. In a thermocouple rf ammeter, the thermocouple junction is mounted in close proximity to a low-value, noninductive resistor, something like 1 ohm. The voltage produced by the thermocouple junction, then, is proportional to the power, or heat dissapated by the resistor. If this resistor is a negligibly small percentage of the load resistance, then the power dissapated in this resistor is proportional to the current flowing in the line.

Most thermocouple rf ammeters of modern design are almost free of frequency effects up to 50 or 60 MHz, although some people argue with good reason that 30 MHz should be the upper limit of confidence in these instruments.

Care must be taken, however, to insure that the meter is installed in a situation similar to the calibration situation. You will frequently see a calibration indication on the face of the thermocouple rf ammeter to the effect "calibrated on a ⅛ inch steel panel." The calibration of the instrument, then, is valid only when the instrument is mounted on the same kind of steel panel. If there is no indication, then the meter was probably calibrated in open air. This means that the meter should be mounted away from any ferrous object, or probably any other metal surface. Most of the time, these meters are mounted on bakelite, plastic, or masonite panels, that are in turn mounted to the steel or aluminum panel of a box or housing. At least one-quarter inch of the insulating material should show between the rim of the meter housing and the hole cut in the metal box or panel.

Most amateur antenna test instruments can be calibrated for high frequency use by comparing the reading of the instrument to that obtained from a good quality rf ammeter. If you construct an rf power meter, for example, a thermocouple ammeter must be used to calibrate the power scale in watts (the meter movement will probably be a DC meter). The following formula below is used.

$$P = I^2R$$

where P is the power in watts, I is the current indicated on the meter in amperes, and R is the resistance of a *noninductive* dummy load resistor, in ohms.

Thermocouple rf ammeters are nonlinear, so we often find more than one is needed. These meters tend to become crowded on the low end of their scales, so the readings cannot be trusted. We may need a 0–5 ampere meter for testing kilowatt linears, and 0–200 mA to test really low-powered devices (less than 2 watts). In general, it is wise to possess several meters with various full-scale ranges between these values.

DUMMY LOADS

A dummy load is not actually an antenna instrument, but is very useful for testing the test equipment to find out how it should work, and is also used to preadjust antenna tuners and other devices before putting them on the air. It is illegal to perform some tests on the air, so a dummy load is needed.

A good dummy load is shielded and will possess a resistance equal to the radiation resistance of the antenna being used—usually 50 or 75 ohms. The resistor(s) used in the dummy load must be noninductive, or the inductance of the resistor will cause a reactive component of impedance. The resistor will then look like a badly out of tune antenna instead of a resonant antenna.

One amateur manufacturer offers a kit dummy load that is suitable for most amateur work up to 150 MHz, and works well up to 30 MHz. But this model obtains a higher power dissipation of 1000 watts by immersing a lower power resistor in motor oil. This is fine, but look out for that oil. It will seep out of the paint can and ruin the floor. Keep the dummy load in a plastic container, or out in the garage between uses.

Alternatively, use a homemade dummy load constructed from noninductive power resistors such as the Dale Electronics NH-100. These could be mounted inside of a shielded box and fitted with rf connectors, such as the SO-239.

Some commercial dummy loads have variable capacitors inside to compensate for frequency effects, so are essentially flat in their response from near DC to 500 MHz. Although ordinarily too expensive for amateur use, these have turned up on the hamfest market with low price tags.

SWR METERS

You may use almost any of the equations in Chapter 15 as the basis for an SWR measurement system. We could, for instance, measure the surge impedance of the transmission line and the feedpoint impedance of the antenna system. The ratio of these two impedances will give us the SWR. But that is a little troublesome.

Alternatively, we could measure the forward and reflected rf power, and then use the power equation. But rf wattmeters and directional couplers are not always available, and are usually too expensive for most amateurs. Of course, if you want to spend around $150, the Bird Electronics Thruline® Model 43 is a good investment for any amateur.

Most low- and medium-cost SWR meters—and that includes almost all instruments purporting to be SWR meters instead of rf wattmeters—use either the voltage or current in the transmission line to indicate the SWR. The idea is to measure the forward and reflected voltages (or currents) and compare them. The advantage of this scheme is that easy calibration in SWR units is possible. In these meters, we would set the forward power meter (or a switch to select forward-reverse if only one meter movement is used) to full scale, and then read the SWR from the reverse meter (or switch position). If the sensitivity of the meter is adjusted to exactly full scale on the forward power position, then the reverse position of the selector will denote the SWR. Simplicity of construction and operation is the reason why these instruments are so popular.

Figure 16-1 shows an SWR meter based on voltage measurements. Note that it is essentially a partial Wheatstone bridge, so it is usually called an *SWR bridge.* Since a 51-ohm resistance is placed in series with the antenna line, this bridge cannot be left in the circuit once the SWR is measured. Versions of the circuit can be left in, however, but these use a 1-ohm series resistance for R3. Diode D1 produces a DC voltage proportional to the forward voltage at the point of measurement, while diode D2 produces a DC voltage proportional to the reflected rf voltage at this point. Switch S1 selects between forward and reflected voltages for display on meter M1. Note that the scale of M1 will usually be calibrated in SWR units, even though it is a DC current meter movement. The oepration of the SWR bridge is as follows:

1. Set S1 to the *forward* position.
2. Adjust the *sensitivity* control (R5) or *SWR ADJ* control (also R5) to full-scale deflection of meter M1.
3. Set S1 to *reverse* and read SWR from the scale.

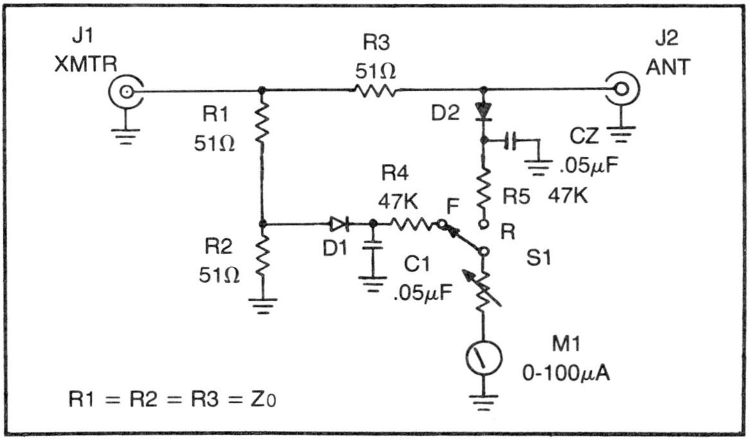

Fig. 16-1. Basic antenna bridge for measuring SWR.

Most of the really low-cost meters of older manufacture use the circuit of Fig. 16-1, but this is not the most desirable circuit because the transmitter power is lost in the bridge. In fact, if more than a couple of watts are used, then the resistors may well burn up. Such instruments then require an external signal source such as a transmitter or small oscillator, rather than the transmitter. Of course, this may be desirable in most cases, even where the SWR meter can be left in-circuit while transmitting. This would allow you to adjust the SWR of the antenna and not interfere with other stations on the air. Use a low-power transistor rf oscillator, crystal controlled, such as the kits sold by International Crystal Manufacturing Company of Oklahoma.

One of the most popular forms of amateur SWR meter is shown schematically in Fig. 16-2. This meter, called the *Monimatch* circuit, enjoys a wide following. Most of the good quality, in-line, SWR meters sold to amateurs (and built by many homebrewers) use this circuit.

The sense element consists of a segment of transmission line (B in Fig. 16-2), parallel and in close proximity to a pair of sense conductors (A and C in the figure). The position of the diodes determines whether forward or reflected currents are measured. Notice that, in both cases, one end of the sense elements is terminated in a resistor, while the other end is terminated in a diode rectifier. The resistor should have a value equal to the expected impedance of the antenna. In most cases, this means either 50 or 75 ohms. But sometimes, amateurs will use a 62- to 68-ohm resistor at these points as a compromise. This value will allow the instrument to be used for both 50- and 75- ohm antenna systems with only a small error (1.4:1 in the worst case). Other than the sensor, this circuit is essentially the same as the previous example, and operation is also the same. Use the instructions for the bridge, just given, for this instrument also.

Construction of the sensor element can be of three methods. The oldest is to modify a piece of coaxial cable. Most of the time, a piece of RG-8 or RG-11 coaxial cable (depending upon which impedance was used for the antenna line) was modified. The outer jacket of insulation was carefully stripped away from a 12-to 18-pinch piece oil the cable. The shield was then exposed. Two small holes were made in the shield at each end of the cable, one on either side of the cable (the holes were diametrically opposed to each other). A length of No. 24 to No. 30 enamel-covered wire was then threaded under the shield in such a way that the ends passed through the holes in the shield, and the two enamel wires were parallel to the center conduc-

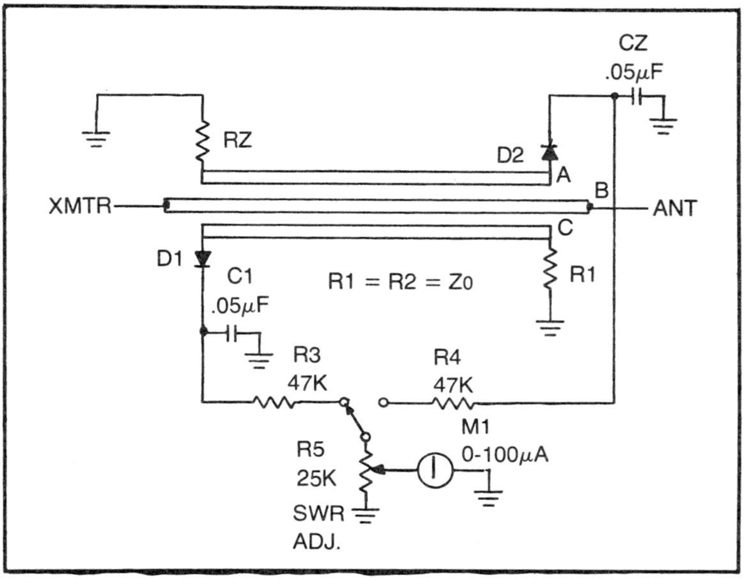

Fig. 16-2. The monimatch SWR sensor.

tor of the coaxial cable. The ends of the enamel wire would be the elements called A and C in Fig. 16-2, while the center conductor of the coax was the element called B.

In more recent times, some construction projects published in amateur magazines showed sensors made from short lengths of 0.125-inch diameter copper or brass hobby tubing, mounted in a plastic assembly drilled to accept the tubing and keep them parallel. In even more recent times, printed circuit boards have been used. The elements A–C are etched into the foil of a piece of copper clad board. In most cases, the length of B would be about 5 inches, and the other two elements a little bit shorter. All of the elements would be 0.25–0.50 inches wide. These dimensions would allow the amateur to use a resist pen to mark the foil, and ordinary ferrous chloride etching solution to make a sensor in under 30 minutes. The only real criteria to worry about is to make the sense elements identical to each other, a chore easily accomplished with a straight edge. It is so easy to *draw* the foil pattern onto the copper surface of the blank board that one should not have to look up the original *QST* reference.

A modified version of the monimatch idea is shown in Fig. 16-3. This is essentially the same circuit, but the sensor has been replaced by a toroid current transformer, T1. The primary of the transformer is the center conductor of the transmission line, which passes

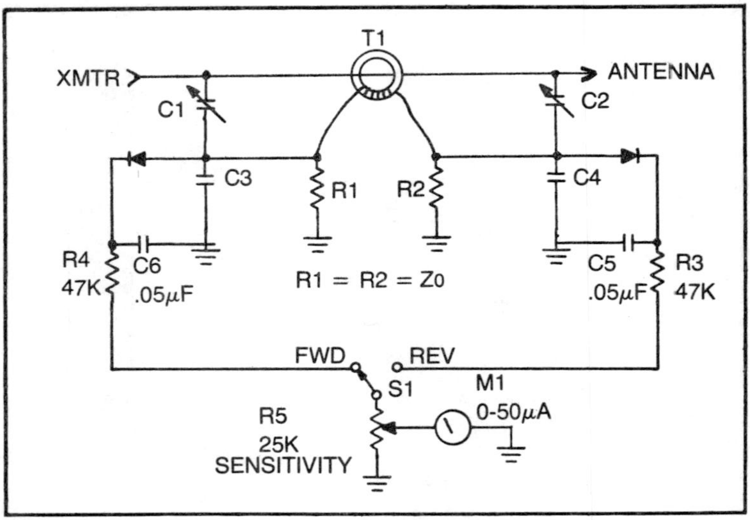

Fig. 16-3. Toroid current transformer SWR sensor.

through the toroid form. A secondary winding of 20–50 turns of No. 26—No. 28 gauge enameled wire is wound onto the toroidal surface. From that point on, the circuit is essentially the same as before, except that only one secondary is used.

The capacitoirs C1–C4 shown in Fig. 16-3 are optional, and are used to frequency compensate the coil for use over the entire high-frequency spectrum. However, the circuit will perform relatively well without these capacitors. The capacitors are, in general, only needed when trying to measure rf power with the instrument. For simple SWR measurements, the device is custom adjusted at each use, so the compensation is not needed. Use 33 pF for C3 and C4 and 7 pF trimmers for C1 and C2. Of course, commercially-made bridges are available (see Fig. 16-4).

ANTENNA IMPEDANCE BRIDGES

An antenna impedance bridge will measure the impedance of the antenna in ohms. Almost all of the bridges used in amateur work are not really true impedance bridges, but measure the radiation resistance of the antenna. Recall that the impedance of any antenna is composed of both resistive and reactive components. If the antenna is perfectly resonant, then the reactive portion is zero, and the total impedance is resistive. In that case, the impedance is the radiation resistance—the rf voltage at the feedpoint divided by the rf current at the feedpoint.

Fig. 16-4. Low-cost bridge using two meters instead of S1.

Most of these simple bridges are based on the Wheatstone bridge circuit of Fig. 16-5. Although the circuit will not measure the reactive component of impedance, a skilled operator can tell whether or not there is a significant reactive component by the way the resistance bridge behaves. A purely resistive, resonant antenna will produce a nice, deep, and sharp null when the variable element of the bridge is adjusted to the same impedance as the antenna (the condition for null). After some experience in this matter is gained the operator can tell by observing the depth and width of the null whether or not a highly reactive component is present.

A representative circuit for an antenna impedance meter is shown in Fig. 16-6. This circuit uses a split-stator variable capacitor for two sections, and the capacitor shaft is connected to a dial calibrated in ohms. The third leg of the bridge is a 68-ohm resistor, a compromise that allows both 50- and 75-ohm antennas to be the design center of the bridge. The antenna forms the fourth, and unknown, leg of the bridge.

Calibration of the bridge is simple. Merely collect a small selection of carbon composition resistors of values between 5 and 1000 ohms. If 5-percent tolerance is good enough for your applications, then merely obtain standard 5-percent resistors. But if closer tolerance is needed, then measure the actual values. Connect each resistor, in its turn, to the bridge and adjust for null. Mark the dial at

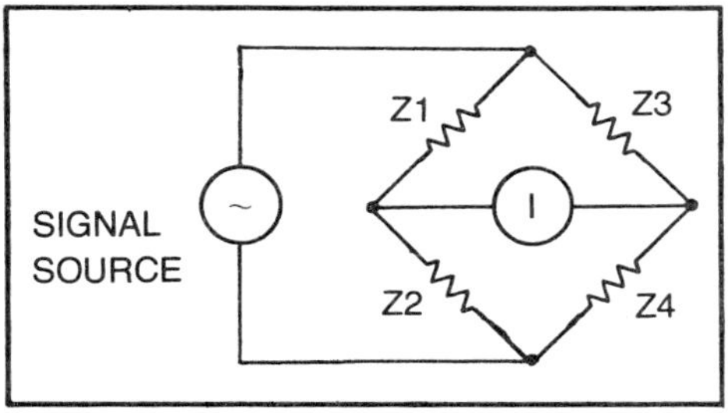

Fig. 16-5. Basic Wheatstone bridge.

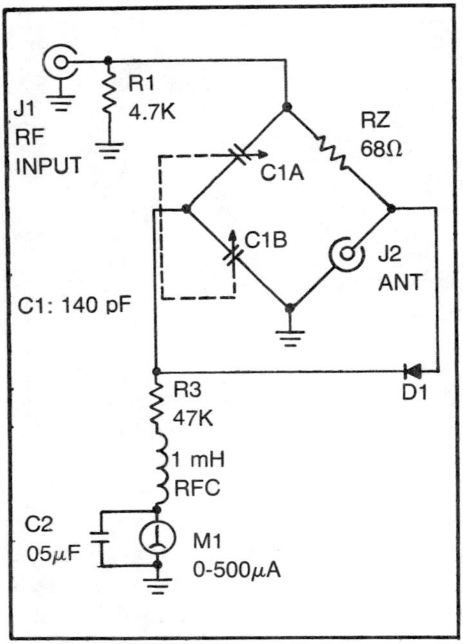

Fig. 16-6. Practical antenna impedance bridge.

the point where the null is obtained with the value of the resistor used for the "antenna." After a sufficient number of points have been calibrated, it is possible to use the impedance bridge for testing amateur antennas. A commercially made antenna impedance bridge appears in Fig. 16-7.

The resistance Wheatstone bridge will not measure reactances. Of course, a resonant antenna is not reactive, so once we have resonated the antenna (see Chapter 18), then only resistance remains in the antenna impedance. Two approaches to measuring the reactive component are used, and these are shown in Figs. 16-8 and 16-9.

In Fig. 16-8, we see the use of a balance inductor and a capacitor. The capacitor will be connected to a shaft that is calibrated in ohms. The capacitor is set to the middle of its range, at which point the dial is marked "0." Inductive and capacitive reactance values, in ohms, are marked on the dial either side of zero. When the reactance dial is set to zero, the inductance of the balance control is adjusted for the best null on the meter. If the null is total, then the antenna is resistive. But if the capacitor must be adjusted to improve the null, then there is a reactive component. The resistive, inductive and capacitive nulls are somewhat interactive, so some practice is needed to learn to use this type of instrument effectively.

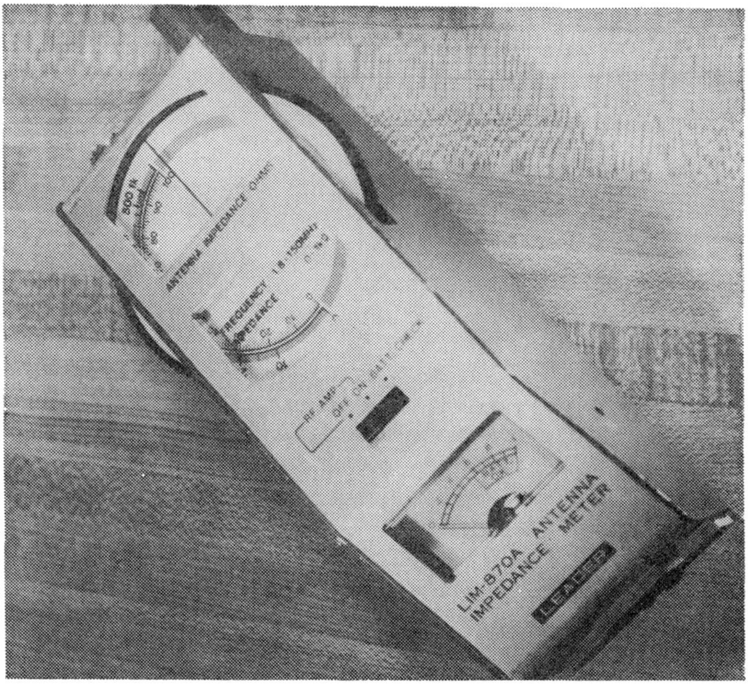

Fig. 16-7. Low-cost commercial antenna impedance bridge measures Z_L from 0–1000 ohms, over the range 1.8–150 MHz.

The second technique is shown in Fig. 16-9. Here we use two capacitors. C2 is a variable capacitor connected to a zero center dial calibrated in ohms. Capacitor C3 has a value equal to the capacitance of C2 at midscale. When the reactance dial is at zero, then C2 = C3. Again, capacitive reactances are indicated on one side of zero, and inductive reactances are on the other side.

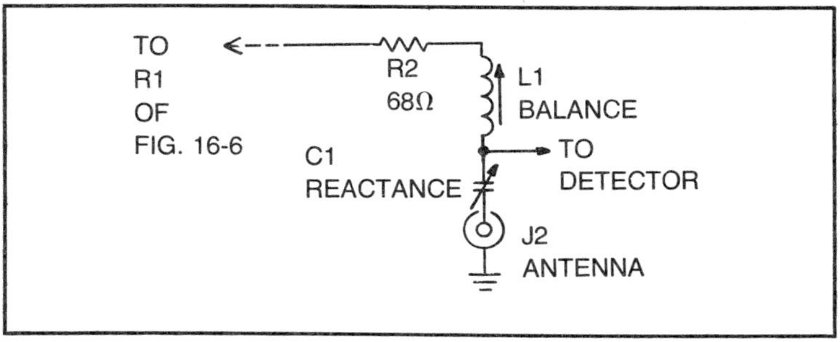

Fig. 16-8. Using a balanced inductor and a capacitor to measure the reactive component.

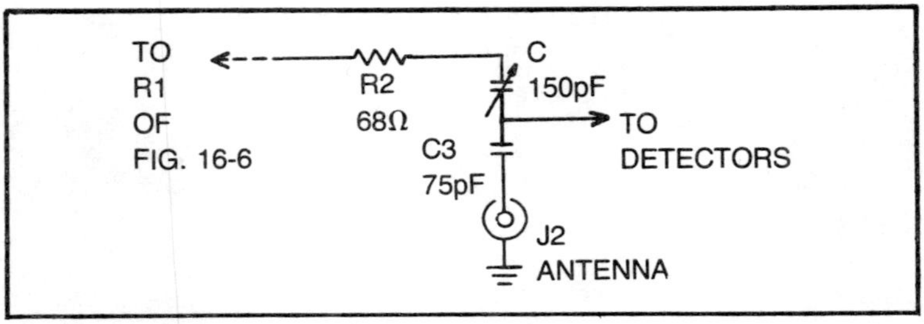

Fig. 16-9. Using two capacitors to measure the reactive component.

NOISE BRIDGES

The rf noise bridge is one of those pieces of test equipment any radio amateur can do without. . . until they use one, after which time the noise bridge becomes indispensible. A knowledgeable amateur would select a good noise bridge over any of the SWR bridges, if only one could be purchased. Of course, it is best to have both types of test equipment, if at all possible.

Figure 16-10 shows the basic circuit of the noise bridge. You can build one of these if you want, but you will find that the price of the Palomar Engineers R-X noise bridge, or the Omega-T bridge (both heavily advertised in amateur radio magazines) are low enough in cost to make it more worthwhile to buy one.

The signal source in the noise bridge is a noise generator. It seems that a reverse biased diode will produce a large amount of white noise, with spectrum components up to 150 or 300 MHz. A special transformer is used in the noise bridge. It is trifilar wound. Coil L1 is used as the primary and is connected to the output of a

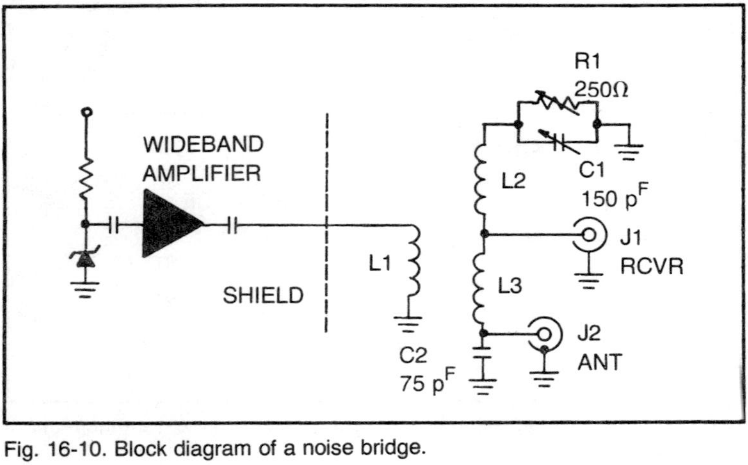

Fig. 16-10. Block diagram of a noise bridge.

Fig. 16-11. Omega-T noise bridge.

wide-band amplifier that amplifies the noise signal. Coils L2 and L3 form secondary windings, and are two legs of the bridge circuit. The other legs of the bridge include R1/C1 as one leg (a calibrated leg) and C2/antenna as the unknown leg. The use of C1 and C2 is the same as for the bridge in Fig. 16-9. In most cases, L1:L2:L3 have a 1:1:1 ratio.

The detector used in a noise bridge is a good general coverage receiver. Amateur receivers can be used too, but only if you can guarantee that all antennas will be cut the first time inside of an amateur band. For most users, the general coverage receiver is

Fig. 16-12. Palomar Engineers noise bridge.

Fig. 16-13. Multipurpose antenna tester that is good to 50 watts rf power.

required (I use an old Hammarlund HQ-145, bought cheap at a hamfest). The receiver must have an S-meter, or at least an agc voltmeter (which you may add, if you are so inclined).

The basic idea of the noise bridge (see Figs. 16-10 and 16-11) is that a high noise level will be heard on the receiver at all frequencies except the one frequency at which the bridge is balanced. When the receiver is tuned to the resonant frequency of the antenna, and the reactance has been nulled out by adjusting capacitor C1, then a noticeable dip in the noise level, as best indicated on the receiver S-meter, will occur. The dip is audible, so you can use your ears, but the effect is easier to see on an S-meter.

There are many applications for the noise bridge, some of which are detailed in Chapter 18. For example, you may use the noise bridge to find and correct the resonant frequency of an antenna, troubleshoot an antenna that has developed problems after working for awhile,find the velocity factor of a piece of transmission line (coaxial cable, mostly), or find the physical length of a piece of coaxial calbe that is an electrical half-wavelength. Want to go the limit? A multipurpose antenna tester is shown in Fig. 16-13.

Chapter 17
Antenna Construction Practices

This chapter will deal with matters relating to the physical construction of radio antenna systems. Our first topic is ordinary soldering. Even if you know how to solder a pair of wires together, you may not be a good solderer of an antenna. The reasons are your tools may not get hot enough and you may get impatient. The wires of an antenna represent a tremendous heat sink. Just because you can solder No. 14 wire with your iron inside, on the bench, does not mean that the same tool will work nicely on the antenna. First of all, the antenna may be 130 feet long, and *that*, my friend, is a large body of metal to sink the heat from your low-power iron. In fact, I once was unable to solder an antenna wire because the iron was a 40-watt printed circuit iron. Only my 200-watt Weller D-440 was able to handle the job. When the wire is long, or when a slight breeze conspires to take away the heat, then a larger iron is needed.

A propane or similar torch may be needed when soldering copper pipe or other large copper objects. In that case, buy unfluxed solder, and a small tin of soldering paste. These are available at most hardware stores. Dip the solder in the paste just before applying it to the heated surface of the pipe. If the pipe is aluminum, then don't attempt to solder it unless you have special equipment. Use mechanical/electrical fasteners for this. These are like cable clamps with a screw that bites into the pipe to insure connection.

WIRE ANTENNAS

Wire antennas can be made very cheaply. Dipoles, long wires, and even verticals (if an insulated support is available at the top and bottom ends) can be made of wire. The best wire is No. 14 or No. 12 steel wire that is coated with an outside layer of copper. This type of wire possesses the stength of steel and the electrical conductivity of copper (remember, because of skin effect, the rf currents flow only on the outer, the copper layer). But Copperweld, as this wire is called, is hard to get. Also, it is expensive. If you have ever worked with Copperweld wire you will know that it will kink in the blink of an eye, and the kink seems permanent. The antenna will seem to have a

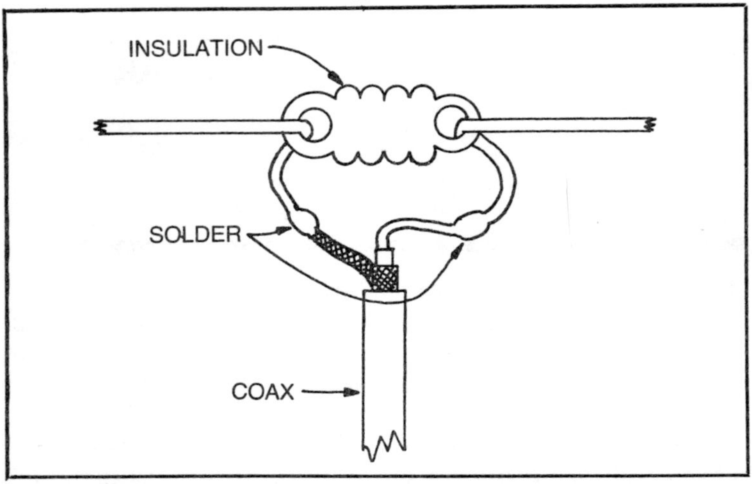

Fig. 17-1. Improper way to make a dipole center insulator connection.

permanent glitch at the point where any kind existed. The next best wire is hard-drawn copper wire, also of No. 14 or No. 12 size. Choose only the stranded wire, because the solid stuff will develop metal fatigue early in the game. That means it will break.

All wire-to-wire junctions of the wire antenna must be soldered. Any exceptions will result in a nonworking antenna in a very short time period. Corrosion will build up and cause the joint to develop electrical resistance. This will waste power, at best, and cause a high SWR that damages the transmitter at worst.

Never depend upon a joint, soldered or unsoldered, for mechanical strength. Always strain relief the joint. It is always best to use a center insulator designed for use in dipoles, rather than an open insulator. Most amateur radio supply houses sell such insulators. They have the advantage of keeping the joints dry during rainy weather. If you *must* use an ordinary end-insulator for the center insulator, however, use the scheme shown in Fig. 17-2. The version in Fig. 17-1 is wrong...and will break in very short order. This is especially true if the transmission line is long and therefore heavy. The coax is wrapped one-turn around the center insulator, and secured with nylon string, thread, or fishing line. The wire forming the radiators is passed through the hole in the insulator and then wrapped over itself to form a strong splice. This gives mechanical strength. The loose end of the wire is then soldered to either the inner conductor or the shielded braid of the coax, depending upon which end it is. The other end of each wire is also terminated in an end insulator and must be treated in the same way. Wrap the wire around itself to obtain strength. On both ends, solder all wire-to-wire contact points. Solder the radiator wires where they are wrapped around themselves at the center and end insulators, and make sure that they are coated with solder. This keeps the junction a low resistance, even after the wire has corroded.

The ends of the antenna may be supported with any type of insulated rope. In most cases, the rope will be tied to each end insulator. The free ends of the two rope supports will then be tied to some support such as a tree, building, or a tower. It is always a good idea to construct wire antennas in the clear, away from any obstructions. This is usually taken to mean at least several wavelengths above the ground, and several wavelengths from any surrounding structures that might interfere with the radiation pattern or change the radiation resistance of the antenna. This means that the rope used to support the antenna might be quite long. Although, keep in mind that most practical antennas on city or suburban lots are not able to meet these nice, ideal specifications.

One popular construction method is to use pulleys on each end of the dipole, with ropes to the ground, so that the antenna can be raised and lowered as needed for trimming of the length during tuning or troubleshooting. One of the simplest ideas is to use a D-ring fitting as in Fig. 17-3. The D-ring is machined from a piece of aluminum or brass, so that it will not corrode. I know several amateurs who have made these, and they claim that they corrode less than pulleys. It seems that even some so-called "marine-grade"

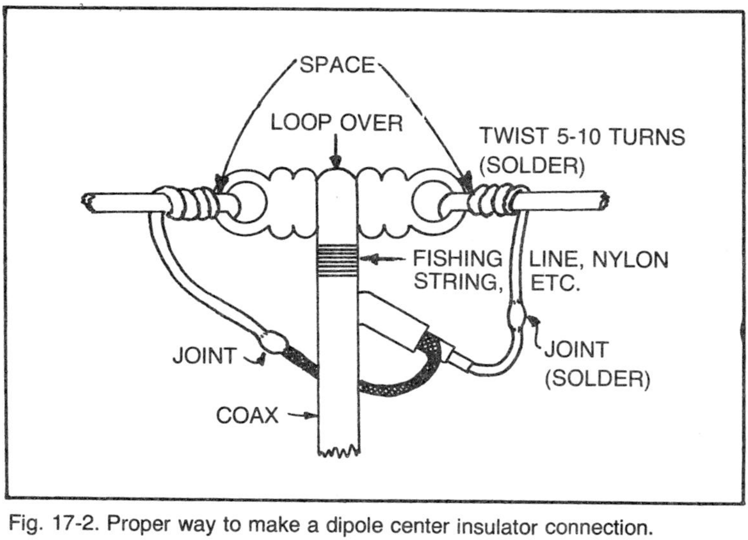

Fig. 17-2. Proper way to make a dipole center insulator connection.

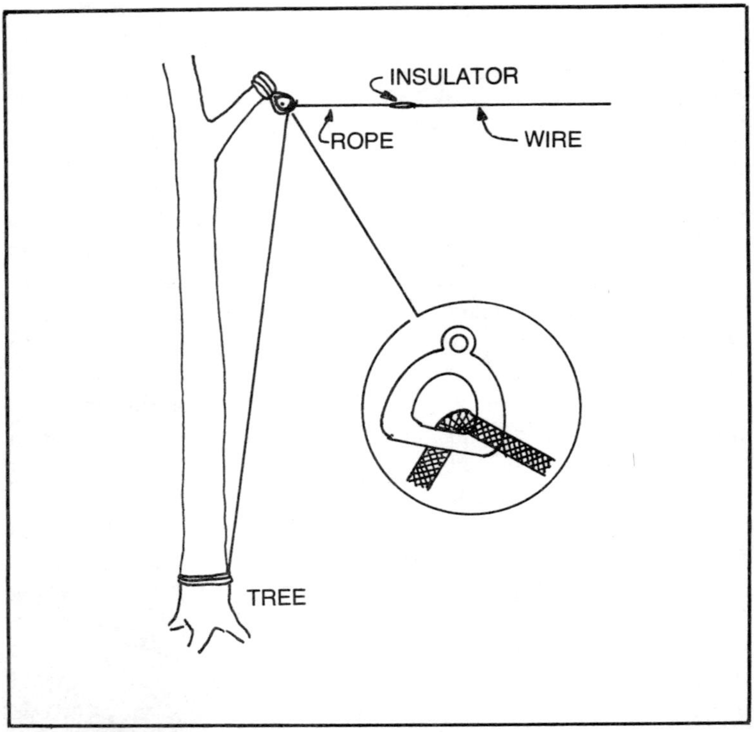

Fig. 17-3. Way to anchor the dipole to a tree so that it can be lowered for tuning and repair.

pulleys have some ferrous metal in them and will corrode in time. Another alternative is to use those marine-grade pulleys, but be careful to get what you pay for.

WARNING!

Under no—repeat *No* circumstances—is it permissible to build an antenna in close proximity to an electrical power line! If the antenna should contact the line while you are working on it, or during a windstorm, then the results might be lethal. Mixing antenna construction with the power mains can certainly be fatal, and that is spelled with a capital F-A-T-A-L, as in dead, killed, cooked goose! Every year we hear about people who either construct a vertical close enough to the lines that it will make contact when it falls over, or who try to throw the wire for a dipole *over* a power line (the idea, I think, is to get the wire above the line—sigh). At least one of these incidents every year is fatal. When the antenna member strikes the AC power wire, the worker is electrocuted.

Chapter 18
Tuning and Troubleshooting Antennas

In both commercial and amateur applications, I have installed, perhaps, hundreds of antennas. Only a very small number, less than two percent I suspect, worked right the first time. Almost all antennas have to be adjusted during installation. The formulas and the tables in this book give a proper starting point. From this starting point you will work toward tuning the antenna for best operation.

TUNING

There are actually two aspects to be considered here. One is resonance and the other is *matching*. Many radio amateurs believe that these are one and the same, but in reality they are merely *related.* An antenna can be perfectly resonant with no reactive component to the feedpoint impedance, yet have a very high SWR. A case in point is a vertical on the 20-meter band which I built a couple of years ago. The antenna length and the radials were trimmed to make the antenna resonant. My noise bridge and Heat dip meter told me so. But there was still a high SWR. Measurement of the antenna impedance, however, told the tale. The feedpoint impedance was on the order of 15 ohms. Since I was feeding the antenna with 50-ohm coaxial cable, the mismatch produced an SWR of (50/15):1, or 3.3:1!

Resonance can be determined using either a noise bridge or a dip meter. The dip meter is a special VFO with the inductor mounted external to the case so that it can be coupled to the circuit being tested. When the oscillator is adjusted to the resonant frequency of the circuit being tested, then energy is transferred between the two circuits, and this causes the meter to dip. Most dip meters are not calibrated well enough to read frequency accurately, so one can improve the results by using a general coverage receiver to listen to the dip meter oscillator signal while making the adjustment.

Figure 18-1 shows the proper method for coupling a dip meter to an antenna. It is possible to couple the meter directly to the antenna by bringing the coil on the meter in close proximity to the antenna radiator. However, this has at least two disadvantages: it is often physically impossible or at least very difficult, such as with a

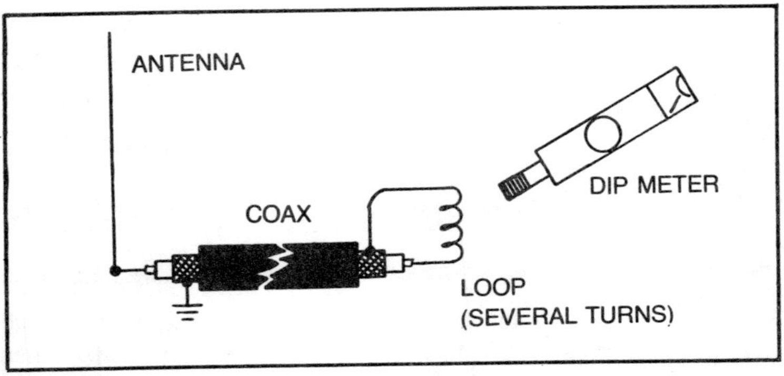

Fig. 18-1. Coupling a dip meter to a coaxial transmission line.

dipole mounted off the ground, and it might well detune the antenna enough to render the data invalid. In Fig. 18-1 we use a length of coaxial cable to get us down on the ground and far enough away from the antenna to reduce the coupling between us and the antenna to a negligible point. This has the advantage of allowing us to use the regular antenna feedline, as built into the circuit to begin with. A few turns of wire are used as a gimmick to couple energy from the dip meter into the antenna circuit. Keep this coupling coil small—only a few turns of wire—so that it does not develop resonances of its own near the antenna frequency . . . thereby obscuring the results.

Using a noise to find the resonant frequency. There are two basic noise bridge designs on the amateur market, those with and those without the reactance capacitors C1 and C2 (Fig. 16-4). Consequently, there are two closely related procedures for finding the resonant frequency of an antenna with a noise bridge.

☐ **When no reactance control is used.**

1. Set R1 to the anticipated radiation resistance of the antenna, (which is normally 50 to 75 ohms).
2. Tune the receiver to the anticipated resonant frequency of the antenna.
3. Adjust the receiver tuning dial very slowly, above and below the anticipated resonant frequency, until a very definite dip is noted in the noise level. Use the receiver S-meter, if available.
4. Adjust R1 for best dip and then readjust the receiver tuning to improve the dip. These adjustments are often interactive, especially on reactive antennas, so some back-and-forth compromising is necessary.
5. When the best null is obtained, read the resonant frequency from the receiver dial.

☐ **When a reactance control is available.**

1. Set R1 to the anticipated radiation resistance of the antenna. Set the receiver to the anticipated resonant frequency.
2. Set the reactance control, C1, to zero (middle of the capacitance range).
3. Adjust C1 for a null either side of X = 0. This is best indicated by a dip on the S-meter.
4. If the null is in the X_C region of the C1 dial, then the antenna is too long. Tune the receiver downband to find the deepest null.
5. If the null is in the X_L region of the C1 dial, then the antenna is too short. Tune the receiver upband to find the resonant frequency, as indicated by a deep dip on the S-meter.
6. Adjust the dial of the receiver, R1, and C1 for the best null.
7. Read the resonant frequency from the receiver dial.

At resonance, the C1 dial will be close to zero, while the R1 dial will indicate the approximate radiation resistance of the antenna. The coaxial cable to the receiver, in both of these adjustments, should be any short, convenient length. For best results, however, the coaxial cable to the antenna should be an integer multiple of a half wavelength. Remember, an electrical one-half wavelength of a piece of coaxial cable is shorter, by the velocity factor of the cable, than a physical half wavelength.

You can also use an SWR meter to find resonance, but I do not like this method too much. The SWR meter can trick you, if you are not careful. The standard instructions are to measure the SWR at the design center frequency of the antenna, and at end ends of the band—the frequencies above and below the design frequency of the antenna. If the SWR improves at the high end of the band, then the antenna is too short. It must be lengthened for best operation. On the other hand, if the SWR improves at the low end of the band, then the antenna is too long and must be shortened. Clip an inch or two at a time. With dipoles, clip an inch or two from each end to keep the antenna symmetrical.

The problem with using SWR is found when the antenna is way off the mark. If, say, a 20-meter antenna is actually resonant at 12–13 MHz (it *can* happen!), then the SWR will not be useful, and you will not be able to find at which end of the band it improves—legally, that is.

The antenna length is adjusted to make the antenna resonant at the proper frequency. If all is well—it won't be—then the SWR will be1:1, and you will be happy. But this is the real world. Antennas are not perfect, and in most cases not even near perfect. The radiation resistance will vary quite a bit from the anticipated resistance. Of course, best SWR and greatest power transfer between the transmission line and the antenna occur when the radiation resistance of the antenna matches the surge impedance of the coaxial cable.

You can use either an impedance bridge or the R1 control on a noise bridge to measure the radiation resistance of the antenna. The impedance bridge may be a little more accurate, but even low-cost noise bridges are calibrated rather well. You may then construct a matching network, or a transformer, to match the antenna impedance. Note that several manufacturers offer antenna transformers. Palomar Engineers, for example, offers a series of antenna transformers designed for just this purpose. Also, you can build your own using a little transformer theory and one of those kilowatt balun kits from Amidon Associates of California.

Using the noise bridge to find a half wavelength of coaxial cable. A piece of coaxial cable is not able to pass a radio signal at the speed of light. As a result, the length of a half-wavelength transmission line is shorter than a half wavelength in free space. We can calculate the approximate physical length to produce a half wavelength of cable *if* we know the velocity factor of the cable. For most foam cables this is 0.80, while for regular polyethylene dielectric coaxial cables it is 0.66. But what do you do if you do not know the velocity factor, or if you want the length more accurately know? Use the noise bridge, of course! The following procedure below allows you to cut a piece of coaxial cable to an electrical half wavelength, using the noise bridge as an indicator.

1. Use the following equation or the tables in Chapter 15 to find the approximate length for the type of coax used, and then cut a length of cable that is five percent longer.
2. Set C1 on the bridge (the X control) to zero.
3. Set R1 to a resistance of a few ohms (just slightly above zero).
4. Connect one end of the coax to the antenna port of the noise bridge, and short the other end of the coax.
5. Tune the receiver first to the frequency at which the cable must be one-half wavelength, and then tune downband *very slowly* until a sharp null is heard.
6. Very slowly, patiently, and a little at a time, cut the end of the cable off, and reshort it. Then return the receiver upband to find the new null.
7. Repeat step 6 until the null has moved upband to the required frequency.

8. Check the result by unshorting the far end of the coax and substituting a 50- or 100-ohm carbon resistor. Then readjust R1 on the noise bridge for best null. Since the impedance of the resistor load will repeat itself every half wavelength, the R1 dial should read the same as the value of the load resistor substituted for the short.

Finding the velocity factor of an unknown piece of coax using the noise bridge. The velocity factor of the transmission line is the percentage of the speed of light at which radio waves are propagated through the line. This may be found from the expression:

$$V = LF/492N$$

where V is the velocity factor, 0 to 1, L is the length of the cable in feet, F is the frequency in megahertz (MHz), N is an integer (1, 2, 3,....).

Procedure:

1. Cut a sample length. It should have a physical length that would be an electrical half wavelength at some frequency well within the range of the receiver (lengths that would resonate near the receiver band edges are no good).
2. Calculate the free-space frequency of the cable using the expression F = 492/L, where the terms are as just defined.
3. Connect one end of the test length of cable to the antenna port of the noise bridge and short circuit the other end.
4. Set C1 =0, and R1 to a value just slightly above zero.
5. Find the F term of the equation by tuning the receiver dial slowly until a null is found. This will probably be found downband from the frequency at which you calculated in step 2.

TROUBLESHOOTING

OK, the antenna has been working for several months, but all of a sudden the transmitter loads funny. Measuring the SWR, you find that it is now 5:1, where it had once been close to 1:1. What to do? First, go inspect the antenna. Look to see if there are any obvious breakages somewhere. If there are, then fix them and go back on the air. If not, some troubleshooting is in order.

The most likely problems will be a defective transmission line—coax does go bad—or a bad solder connection or metal-to-metal screw connection on the antenna. If the antenna is a dipole or other wire antenna, then it will possibly have to be lowered to the ground in order to troubleshoot the problem. One neat trick is to disconnect the feedline from the antenna and then connect a noninductive dummy load resistor in place of the antenna, or terminate the coaxial cable in its surge impedance. If the signal source is a noise bridge or a signal generator/oscillator, then a 2-watt carbon resistor will work well. Go measure the SWR. If it is at all reasonable, something less than 2:1, then the problem is probably in the antenna. If the problem still exists, then it is in the coax, so replace it. If the problem proves to be in the antenna, then suspect *connections.* In wire antennas this means solder joints, which go bad by the dozen. In pipe antennas (verticals, beams, and such), a mechanical junction has probably corroded. Clean and tighten the joint, or resolder the wire.

Chapter 19
Antenna Matching Systems

Many real-world antennas will require some form of impedance matching before they can be used. Rarely is the 72 ohms predicted for the half-wavelength dipole, or the 50 ohms predicted for the quarter-wavelength vertical actually found in practice. There are almost as many schemes for matching antennas as there are antennas. In this chapter, we will consider a few of them.

L-SECTION TUNERS

Probably the simplest of the tuned matching systems is the L-section coupler of Figure 19-1. In this circuit, we have a series inductor and a shunt capacitor, forming an L-network. The capacitor and the inductor are variable, and are adjusted for minimum SWR at the frequency of interest using an SWR meter. Some amateurs have built rather elegant L-section couplers in shielded boxes, using Johnson roller inductors for L1 and a transmitting variable for C1. Note well that C1 must be a widespaced, transmitting variable because of the high voltage encountered at this high-impedance end of the tuner.

PI-SECTION COUPLERS

A pi-section coupler, as in Fig. 19-2, is a variation on the L-section idea. In this case, both the input and the output of the tuner is resonated. A BASIC program for calculating the component values of an pi-section coupler is given in TAB book No. 1000, *57 Practical Programs & Games in BASIC*.

LINE TUNERS

A pair of line tuners for coaxial transmission lines are shown in Figs. 19-3 and 19-4. The version in Fig. 19-3 uses a pair of series-connected inductors (up to 30 turns of Air Dux coil stock on 160 meters and proportionally fewer turns on higher bands). Capacitor Cl will be between 30 pF on the higher bands, and 1100 pF on the lower bands. In some cases, a multi-section broadcast receiver

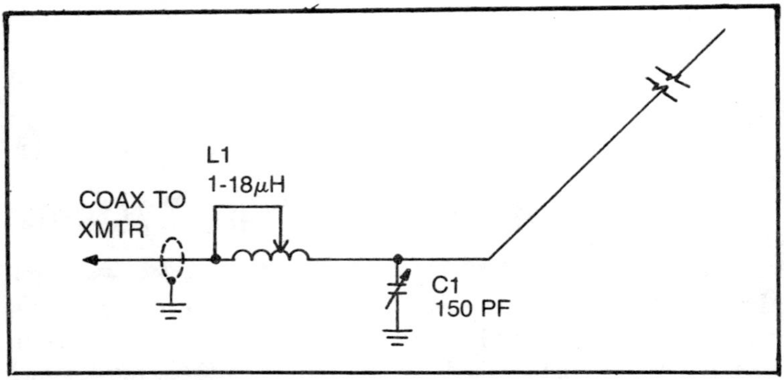

Fig. 19-1. L-section coupler.

variable capacitor is used for Cl (the operating voltages are low because this is a coaxial cable system). The individual sections are wired into the circuit, or switched in if you have a real rf switch, as needed. In general, we use L2 to resonate the antenna, and then L1/C1 matches the input of L2 to the transmission line.

A more popular circuit is shown in Fig. 19-4. This is a wider range line tuner. Again, we are matching coax to coax. The capacitors need not be so-called transmitting variables, as the voltages are low. With up to 200 watts, the ordinary receiver-type variables can be used.

Capacitor C1 in Fig. 19-4 is a split-stator capacitor, meaning that two capacitors share the same frame. It must be mounted so that its shaft is insulated from ground. The shaft, it seems, controls the rotor, which is the point at which the coax from the transmitter is installed. The inductor may be either a Johnson roller inductor or a switched piece of Air Dux or B&W coil stock.

MATCHING BALANCED FEEDERS

Some antennas are fed with balanced transmission line. Examples of balanced line include 300-ohm twinlead and ladder line

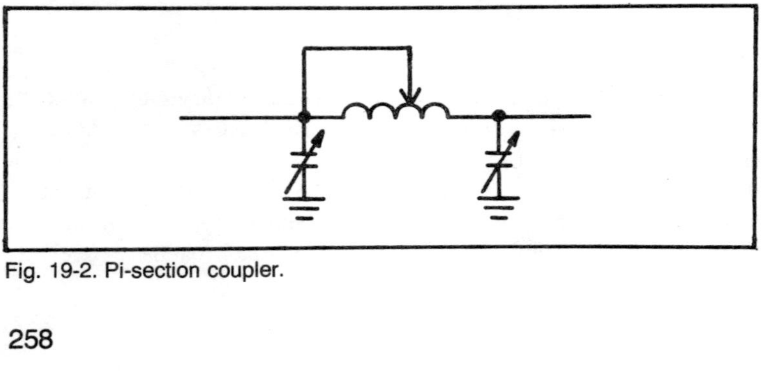
Fig. 19-2. Pi-section coupler.

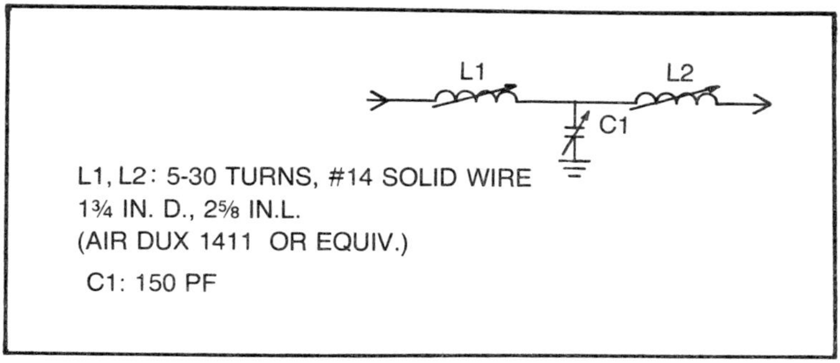

Fig. 19-3. Inductor line tuner.

which is an open type of twinlead. The tuner in Fig. 19-5 will match open line and twinlead.

The primary of the transformer formed by L1/L2 is tuned by capacitor C1. This tunes the coaxial cable end of the tuner. A split-stator capacitor, C2, tunes the antenna side of the tuner. Taps from the secondary coil, L2, are selected by alligator clip leads. Usually, the same number of turns are selected on each side of center.

As with previous types of tuners, the balanced tuner is adjusted for minimum SWR on an SWR meter placed in the coax side of the system. In all of these, incidentally, the SWR meter is on the input side of the tuner.

MATCHING VERTICALS AND MOBILE ANTENNAS

Mobile antennas, and short verticals, must be matched using a circuit similar to Fig. 19-6. In this case L1, and possibly C2, are used to resonate the antenna. A tap on L2 is then selected so that the coaxial cable looks into an impedance equal to its surge impedance. Capacitor C1 is optional, but some mobilers use it to tune the antenna remotely. It is often easier to obtain variable capacitors than variable inductors, so they will select a tap on L1 for the band of operation, and then motor-tune C1 for the specific frequency.

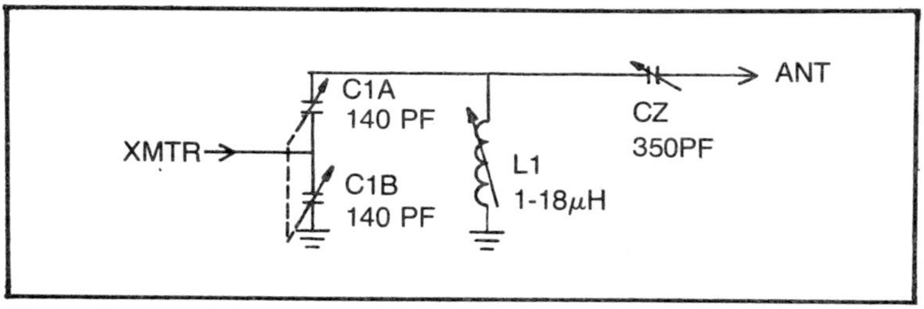

Fig. 19-4. General line tuner.

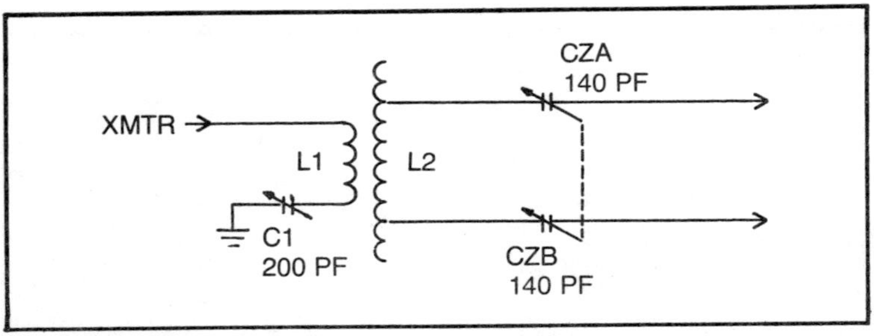

Fig. 19-5. Line tuner for balanced transmission lines.

TRANSFORMERS, BALUNS, ETC.

More and more amateurs are using baluns and wide-band transformers to effect the matching of their antennas. You can now buy inexpensive 1:1 and 4:1 balun (*bal*anced:*un*balanced) transformers. These will match the impedance of an antenna 4:1. In most cases, this means a 300-ohm folded dipole will match 75-ohm coaxial cable. The 1:1 balun is used to give the antenna balanced feed, a measure said to reduce TVI caused by transmission line radiation, and render the signal a little bit better. Do not attempt to use a balun on the same antenna where an antenna tuner or matchbox is used. It will detune the tuner and cause a high SWR. Either the balun or the transmitter may be damaged.

You can also buy wide-range antenna transformers intended to match almost any impedance in the 5- to 75-ohm range to either 50- or 75-ohm coaxial cable. Palomar Engineers offers a series of these transformers.

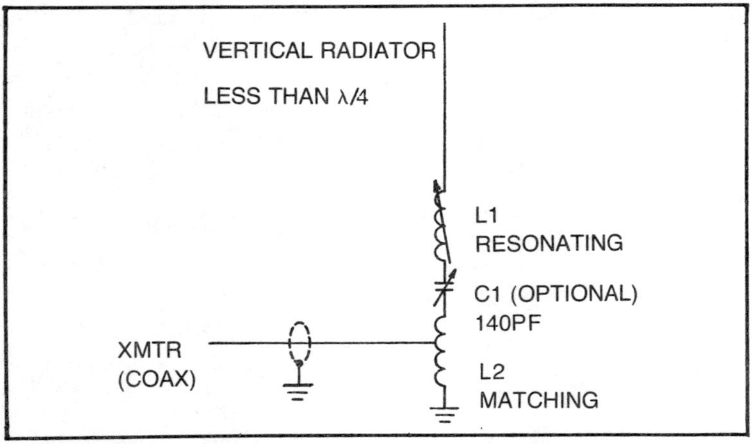

Fig. 19-6. Short vertical antenna, such as used by mobiles.

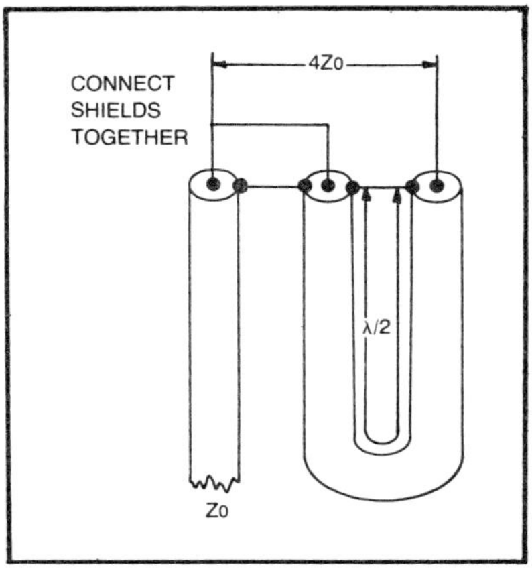

Fig. 19-7. Coaxial cable Balun transformer.

Figures 19-7 and 8 show how to make a single-band balun from coaxial cable. A matching section of coaxial cable, exactly one electrical half wavelength long, is used. The output of this transformer wants to look into an impedance equal to four times the surge impedance of the coaxial cable.

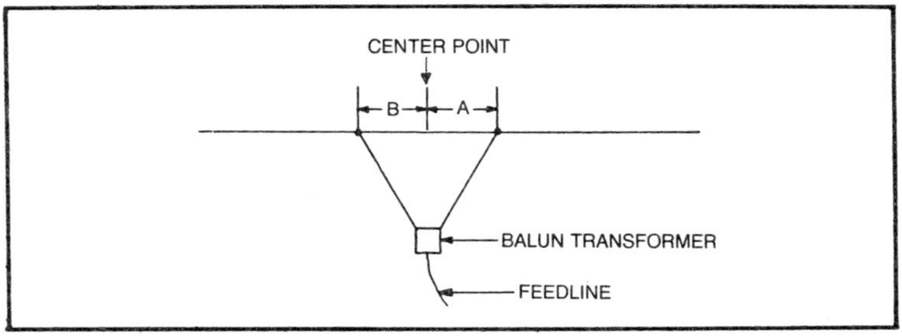

Fig. 19-8. Using a 4:1 balun to feed a half-wavelength dipole.

One use for this type of balun is to delta feed a half wavelength dipole. Ordinarily, the dipole is broken in the exact center, and is fed with 75-ohm coaxial cable. But if the antenna is not a good match for 75-ohm cable—a likely situation in many practical antennas—then delta feeding may be in order. The balun can be constructed of either 75- or 50-ohm coaxial cable. It doesn't really matter which is used, because the feedpoints can be adjusted to either 200 or 300 ohms, as needed for the two types of cable (remember $Z = 4Z_0$). It is important that the symmetry be maintained about the center point. In other words, make length A equal to length B. Adjust the feedpoints a little at a time, while monitoring the VSWR, and then adjust for minimum SWR.

Chapter 20 Limited-Space and Hidden Antennas

Any amateur radio operator who lives in an apartment, boarding house, or townhouse (with a pernicious homeowners association) may want to use a limited space or hidden antenna. You may also have to use a limited space antenna if your lot is small, and the next door neighbor is not too sympathetic to amateur radio activities. . . "keep it on your own side, bub!"

The various types of verticals are among the best limited space antennas. They require only a few square feet of yard or rooftop space, because most of the antenna is straight up in the air. If limited space is the problem, then you may well consider a quarter wavelength vertical as the antenna of choice. These antennas work well on 20 meters and up.

The dipole can also be a limited space antenna. This can be taken in two ways. One is that the antenna run need not be perfect, and some nearby objects can be tolerated. This is especially true if matching problems can be cured. In another sense, we can call the dipole a limited space antenna when it is installed in a nonconventional manner.

Figure 20-1 shows a slanted dipole, or *slipole or sloper,* antenna. One end is supported high off the ground by a 2×4 at the top of the house, while the other is supported by a stake driven into the ground (or in one of my antennas, the bottom rung of a picket fence). The feedline should be brought away from the antenna at an angle as nearly 90 degrees as possible. This cannot always be guaranteed, of course, and if not possible will not represent too great a loss.

Another idea for the dipole is to bend the ends to accommodate the site, as in Fig. 20-2. Funnily enough, Fig. 20-2 could represent either a top or side view: it doesn't matter which way you bend the ends. In most cases, however, we would consider Fig. 20-2 as being a view from above. If possible, make sections A and B the same length, and C and D the same length. Also, see to it that A and B are longer than C and D. Again, this is the best situation, but if not possible, why not try the antenna anyway? The angle between the sections, θ, should be 90 degrees or less. Of course, if it is 0 degrees, then we would have a regular dipole.

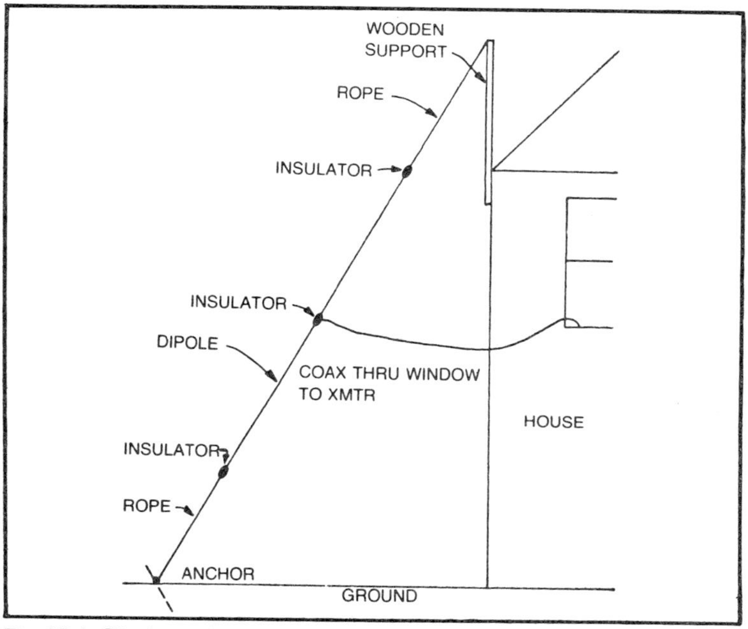

Fig. 20-1. Slipole or sloper antenna for limited space.

Figure 20-3 shows another version of the bent dipole. This is the version normally found when the dipole must be deviously hidden from view, lest the scrutiny of the local homeowners association fall on your castle! Such an antenna could be installed in an attic, or even

Fig. 20-2. Z-dipole.

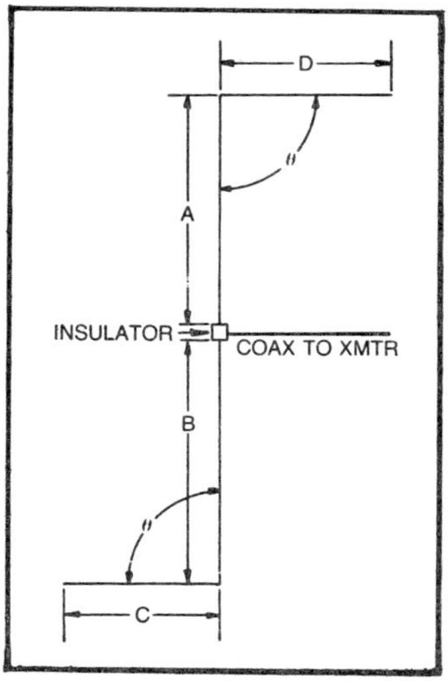

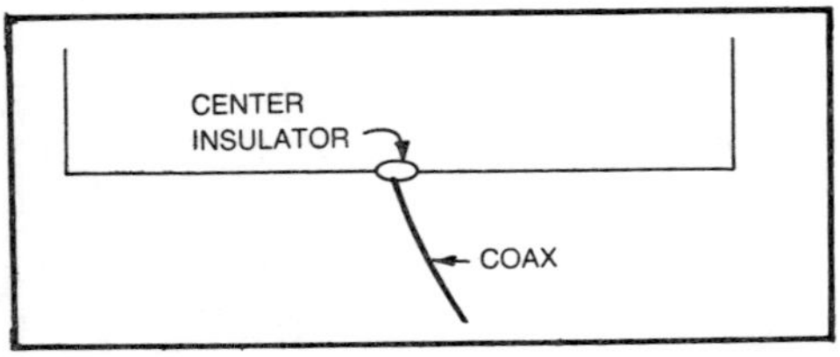

Fig. 20-3. U-dipole.

in a room. When I was in college, I had several 15-meter dipoles taped to the ceiling of my boarding house room so the landlady wouldn't know. . .well, unless TVI became a problem.

Two very popular portable or hidden antenna designs are shown in Figs. 20-4, and 20-5. These are actually little more than versions of the same idea. If you mention the need for a portable or hidden antenna to some people, you will inevitably hear the same tired solution: use a long wire. The instructions are to run a wire of at least a quarter wavelength or more out to a tree or another building. Make the wire No. 30 so that the landlord will not see it. Never mind that it will break the first time you have a 10-knot gust. We are then told to connect an antenna tuner to the radiator wire and a *good ground*. They always talk about the good ground. But usually, the very circumstances that prohibit us from building a really proper antenna also prevent us from running the long wire and from obtaining that good ground (sigh). Tuning a long wire, or almost any other antenna that requires an extensive matching network, does require good ground. In fact, let's reiterate that for emphasis: it needs a *good ground.* I have found that using a long, insufficiently earthed, ground wire will make tuning the long wire antenna a lot like trying to nail Jello to the wall!

But Figs 20-4 and 20-5 show two reasonable alternatives. In Fig. 20-4 we see a pipe antenna. The physical construction should be such that it can be taken apart when not in use, and stored away as a

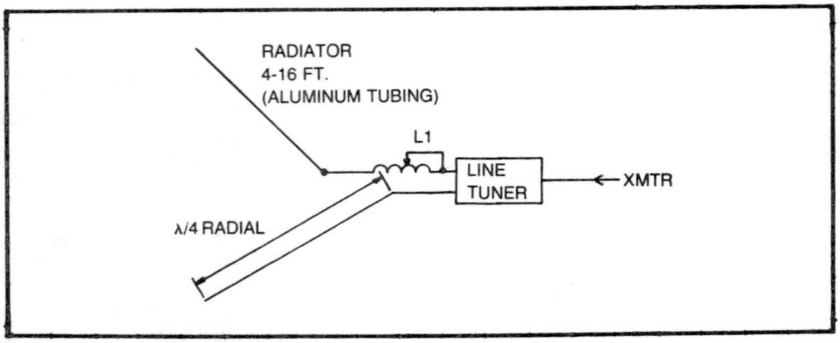

Fig. 20-4. Out-the-window portable, hidden antenna.

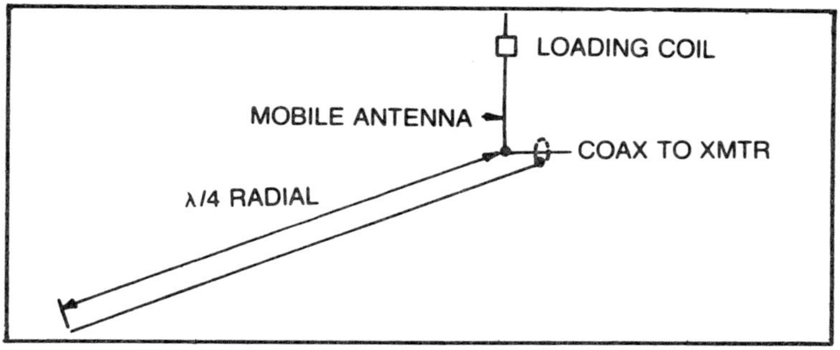

Fig. 20-5. Using a mobile antenna for same purpose.

portable antenna. It should also be built so that it can be easily set in the open window of your operating place. The coil is one of those B&W miniductors. Choose one with a diameter of 1.5–3.0 inches. Usually only a few turns are needed. Using the 2-inch stuff, with 12 turns per inch, I have had good results on 20-meters with a 10-foot aluminum pole; in fact, only five turns were needed. An alligator clip lead, or rf switch, will select the correct number of turns. This will vary markedly from one end of the band to the other. An antenna line tuner will probably also be needed to match the coax of the antenna. The coil resonates it, but does not match it.

Another similar alternative is shown in Fig. 20-5. In this case, we are using a standard mobile antenna such as the *Hustler R* series. The antenna will be mounted in the same manner as in the previous example, so that it will be outside a window during operation and pulled inside when not in use.

Both of these antennas depend very heavily on the quarter wavelength radial. It is mandatory! If you cannot leave a quarter wavelength radial permanently installed outside, then nail it to the baseboard inside the room. I did this, and it worked pretty well. Of course, it is best if installed outside. If this is impossible however, it will work a lot better than some poor ground. The power line ground in the house is often many wavelengths long.

Will these antennas work as well as a dipole or a vertical? In a word: NO. They will not work as well as a proper antenna. But their intent is not working as well as a proper antenna. It is to get you on the air under circumstances where it might otherwise be impossible, or at least terribly difficult.

Index